Methods in Plant Pathology Volume 2

Methods for the Diagnosis of Bacterial Diseases of Plants

R.A. LELLIOTT MA
& D.E. STEAD BSc PhD DIC
Agriculture Development & Advisory Service,
Ministry of Agriculture, Fisheries and Food,
Hatching Green, Harpenden,
Hertfordshire

Published on behalf of the
British Society for Plant Pathology by
Blackwell Scientific Publications
Oxford London Edinburgh
Boston Palo Alto Melbourne

Editorial offices:
Osney Mead, Oxford OX2 0EL
8 John Street, London WC1N 2ES
23 Ainslie Place, Edinburgh EH3 6AJ
52 Beacon Street, Boston
Massachusetts 02108, USA
667 Lytton Avenue, Palo Alto
California 94301, USA
107 Barry Street, Carlton
Victoria 3053, Australia

First published 1987

Set by Setrite Typesetters
Printed and bound in Great Britain by
William Clowes Limited,
Beccles and London

DISTRIBUTORS

USA and Canada
Blackwell Scientific Publications Inc
PO Box 50009, Palo Alto
California 94303

Australia
Blackwell Scientific Publications
(Australia) Pty Ltd
107 Barry Street
Carlton, Victoria 3053

British Library
Cataloguing in Publication Data

Lelliott, R.A.
Methods for the diagnosis of bacterial diseases of plants. — (Methods in plant pathology, ISSN 0266-5514; v. 2)
1. Bacterial diseases of plants — Diagnosis
I. Title II. Stead, D.E. III. British Society for Plant Pathology IV. Series
632′.32 SB734

ISBN 0-632-01233-1

Library of Congress
Cataloging in Publication Data

Lelliott, R.A. (Ronald A.)
Methods for the diagnosis of bacterial diseases of plants.

(Methods in plant pathology; v. 2)
Bibliography: p.
Includes index.
1. Bacterial diseases of plants — Diagnosis. 2. Bacteria, Phytopathogenic — Identification. I. Stead, D.E. (David E.) II. British Society for Plant Pathology. III. Title. IV. Series.
SB734.L45 1987 632′.32
86-28407
ISBN 0-632-01233-1

Contents

Foreword

This book is intended mainly for those in extension services who need to identify bacterial plant diseases rapidly so that the farmers and growers they advise can act in time to avoid or reduce crop loss. It should also be useful to those doing research on such diseases, who need means of distinguishing the pathogen they are interested in from the many other bacteria in or on plant material or in the environment; and of course for those who teach these workers. It is based on the collective experience at this Laboratory, going back to 1951, in diagnosing bacterial diseases of UK crops for the extension service of the UK Ministry of Agriculture, Fisheries and Food (MAFF), as well as those on plants or plant parts imported from overseas and intercepted in the UK.

One of the Laboratory's jobs over these years has been to develop a diagnostic methodology for such plant bacteria for use by the Regional Laboratories of the extension service and for its own purposes. As curators of the National Collection of Plant Pathogenic Bacteria we have also had to develop and refine methods for confirming easily and readily the identity of cultures in that Collection (whose 3000 cultures cover most of the world's plant pathogens) both when they are deposited in and are handled by the Collection. This has helped in identifying key characteristics for determination of overseas pathogens.

We feel that these methods have stood the test of time long enough to be recommended to others outside our own organization. Inevitably those for the diagnosis of pathogens that occur in the UK have been more rigorously tested than those for pathogens alien to the UK, but no methods are recommended in this book that have not been tested or used by us.

It would be hurtful to those overlooked to try to acknowledge all the members and former members of MAFF who have contributed to the development or testing of methods given here. But it would be unfair on this count not to acknowledge the outstanding contribution of Lilian Baker, formerly of the Wye sub-centre of SE Region, Frank Catton, formerly of Eastern Regional Centre, Cambridge, and Jennifer Sellwood of, and Pauline Roberts formerly of Harpenden Laboratory.

Ronald Lelliott, David Stead
Harpenden Laboratory

1

General Aspects

1.1 Introduction

Some 25 years ago, one of us (R.A.L.) produced a manual of methods for the diagnosis of bacterial plant pathogens found in the UK. This manual was issued for internal use within the extension service of MAFF and was intended to provide the organizations in the service with the methods in use at what was then the Ministry's Plant Pathology Laboratory at Harpenden and now the Harpenden Laboratory of ADAS Agricultural Science Service. As a result, the methods were used extensively and routinely at some ten Regional Laboratories of ADAS, often by technicians, and any that were not robust enough to give consistent and unambiguous results in practice were discarded. Over the years since, new methods, as they have been developed, have been rigorously screened in this way and, if successful, incorporated into the manual.

The methods have served two objectives that can, conveniently, be termed presumptive and confirmatory diagnosis. Presumptive diagnosis is to meet the need, usual in advisory work, to forego the rigid proof of the identity of the pathogen so that advice can be given to the farmer or grower within a few days. A presumptive diagnosis based on symptomatology and either the microscopic examination of tissue and recognition of the pathogen on isolation media and/or, for certain diseases, 'phage, serological or a few rapid biochemical tests is usually sufficiently accurate to allow such advice to be given.

Confirmation of such presumptive diagnoses is intended, where time allows, to give workers inexperienced in these groups of bacteria confidence in their diagnostic ability and provide an accurate standard against which plant pathologists can check their visual diagnoses; confirmation would obviously also be necessary where highly accurate results were more important than rapid answers.

This book is based on that manual, but its scope widened to include most known bacterial pathogens of plants and not just those occurring in Great Britain; it also has more detailed descriptions of symptomatology. It does not cover the streptomycetes, certain fastidious prokaryotes such as spiroplasmas and xylem- and phloem-limited, rickettsia-like organisms that cannot be grown artificially or require highly specialized media, and of which the authors have had no direct experience. It also includes not only those well-tried and proven methods used in ADAS for pathogens in the UK but a number which have not been so tested but for which there appear reasonable grounds (often based on our experience in the National Collection of Plant Pathogenic Bacteria of handling pathogens non-indigenous to the UK) for expecting them to be reliable and useful in routine diagnostic laboraties. It is not intended to cover the taxonomic methods used in classification. Essentially it combines description of the older, traditional diagnostic methods with those newer methods that can readily be used routinely.

Although the newer nomenclature for the Gram-positive pathogens has been adopted in this book, the general term 'coryneforms' has been retained to include all those formerly in the genus *Corynebacterium* and now in *Clavibacter*, *Curtobacterium* and *Rhodococcus*.

1.2 Selection and treatment of specimens

Specimens of plants, or parts of plants, should be selected to reflect the range of symptoms in the crop. This is imperative because symptomatology is the starting point for diagnosis and because more than one pathogen or disorder may be present. It is also important that specimens selected include material in a state from which isolation of the primary pathogen can be readily made or will at least be possible. Pathogens are more readily isolated from the earlier stages of disease. Advanced diseased tissue will often contain saprophytic micro-organisms which grow on tissue killed or debilitated by the primary pathogen and which will outgrow the pathogen on non-selective isolation media. Thus the earlier stages of the disease should always be included if present when specimens are taken from a crop. Specimens of cankers and shoot dieback should include the edge of the lesion and a few inches of the healthy material beyond it. The collection of typical healthy material at the same stage of growth as that diseased is often helpful for later comparison with diseased material in the diagnostic laboratory.

Specimens of most diseased material intended for laboratory diagnosis should be placed in polythene bags without delay, should be free of surface moisture and should reach the laboratory as quickly as is feasible. They should be kept as cool as possible during transit and not left in hot, parked vehicles. If unwashed roots form part of the specimen they should be prevented from soiling the rest of the specimen by tying a separate polythene bag around the roots and base of the stem. Soft rot specimens should never be placed in polythene bags but should be packed in a box with crumpled paper or wrapped in absorbent paper.

Microscopic examination and, where necessary, isolation of the pathogen on artificial media, should be done as soon as possible. Where immediate examination is not possible, specimens may be stored in a refrigerator for short periods in order to delay secondary invasion of diseased tissue by other micro-organisms. Specimens should be kept (at 2–5°C) after isolation has been attempted in case further isolation attempts are required, although it is often preferable to obtain further specimens of fresh, young, infected material if this need arises.

1.2.1 Leaf spots and necroses

Necrotic areas usually arise from extension or coalescence of leaf spots. Leaves with small leaf spots should be included if present and should be looked for on or around those plants showing necrotic leaves. Necrotic areas are generally poor sites for isolating pathogens since secondary bacteria have often invaded the necrosed tissue.

1.2.2 Fruit spots

Selection should be as for leaf spots.

1.2.3 Wilted leaves and stems

Wilting is often due to blockage of the plant's xylem vessels; consequently the pathogen may not be present throughout the part of the plant in which symptoms are apparent. Whole plants should therefore be selected when possible.

1.2.4 Galls

Young galls should be selected. Care must be taken not to damage root galls which may become contaminated with soil bacteria.

1.2.5 Cankers and shoot dieback

Specimens should always include the edge of the lesion and several centimetres of healthy material beyond it since the pathogen will usually be most numerous at the edge of the lesion but absent or in low number in older canker tissue.

1.2.6 Soft rots

It is essential that specimens are not in an advanced state of rotting. Fleshy tissues such as leaves and tubers rot rapidly and treatment of specimens should allow for this. Isolation should be made as quickly as possible to reduce the chance of secondary invaders succeeding the populations of the primary pathogen. Specimens should reach the laboratory without delay.

1.3 Microscopic examination

Careful microscopic examination is an essential part of diagnosis. Suitable material should be mounted on a flamed slide in a drop of sterile water to facilitate subsequent isolation (1.4). A dissecting stereoscopic microscope is useful when preparing material for examination with a high-power microscope. Leaf and stem spots and the advancing edge of the lesions of leaf and stem necroses should be excised and bisected with fine scalpels or needle-knives; many leaf spotting pathogens produce a toxin which causes a halo of chlorotic tissue, free of bacteria, round the dark, necrosed part of the lesion and it is important to cut through this necrosed tissue. The primary pathogen can usually be found in large numbers when material from the advancing edge of canker lesions and discoloured vascular tissue is teased out. Teased or squashed material from a millimetre or so behind the edge of the lesion usually yields soft rot pathogens in large numbers. The bacteria in hyperplasias (galls, etc.) are usually few and can only be demonstrated visually by special techniques.

The use of phase contrast microscopy is a great advantage when differentiating bacteria from cell debris but ordinary, dry, × 40 and × 100 objectives can be used if the aperture diaphragm is stopped down. If there is difficulty in such differentiation it can usually be resolved by smearing a loopful of the suspension from the slide onto another slide, staining with methylene blue (6.1.1) and examining under oil-immersion at about × 1000 magnification.

With both leaf spots and cankers, it is advisable not to comminute material but rather to make clean cuts through small pieces of diseased tissue in water on the slide so that the cut edge can be examined. If the cause of an active lesion is bacterial, a cloud of bacteria will usually stream from the cut edge which, particularly under phase contrast, is readily differentiated, with experience, from cell debris. In the case of vascular diseases the vessels should be teased out by holding a piece of excised vascular tissue with one needle-knife and gently scraping the tissue with the other using a dissecting microscope to observe that the vessels are being separated; in some diseases, e.g. slow wilt of carnations, the pathogen in the xylem vessels is embedded in a polysaccharide matrix which is exposed when the vessels are broken and their ends drawn apart. This can be useful diagnostically.

With few exceptions, failure to detect large numbers of bacteria in diseased material indicates either that the disease is not caused by bacteria or that the wrong part of the specimen has been examined. If fungal hyphae or nematodes are seen, it is likely that a fungal or nematode pathogen but not a bacterial pathogen is the primary cause, that these are secondary organisms or that the diseased material has been invaded by saprophytic bacteria. In such cases other lesions and parts of lesions should be examined. Unless diseased material (other than hyperplasias) has large numbers of easily detectable bacteria without hyphae or nematodes, a non-bacterial agent of disease should be suspected.

1.4 Isolation procedures

Most plant pathogenic bacteria grow well on a wide variety of relatively simple media. The more fastidious prokaryotic pathogens — the mycoplasmas, spiroplasmas and xylem- and phloem-limited rickettsia-like bacteria — either do not grow on artificial media or require highly specialized media, and they have not been included in this book.

Although many selective media have been recommended for the isolation and identification of plant pathogenic bacteria, in practice many have not the selective value claimed for them and some are toxic to the pathogen(s) for which they are intended to select; those whose value we have not confirmed have been ignored in this text. Since most plant bacteria grow well on any of a few simple media, often producing characteristic colonies, selective media are usually of use only where plant pathogenic bacteria are to be isolated from very mixed microbial communities such as soil or decomposing plant debris, or where comparatively few cells of the pathogen but many cells of saprophytes are likely to be present in diseased tissue, as is the case with old gall tissue.

A preliminary presumptive diagnosis can usually be made from symptoms and this will influence the choice of isolation media. Suggestions for such media are given in Chapter 3. For some relatively fastidious bacteria, e.g. *Clavibacter michiganense* subsp. *sepedonicum, Erwinia tracheiphila, Xanthomonas populi*, more nutritionally exacting media are required. These are described in the relevant sections.

Generally, isolation should be attempted from material in which large numbers of bacteria have been demonstrated by microscopic examination (see 1.3 above). If slides and thin coverslips (both of heat-resistant glass) and instruments are sterilized by alcohol flaming, and sterile water is used for this examination, and if tissue is then further teased apart to release more bacteria, isolation plates can be streaked with a loopful of suspended material after the coverslip is removed. It is often difficult to release the pathogenic bacteria from gall tissue, and it is necessary in this and a few other cases to comminute tissue in a few drops of water in a sterile Petri dish and leave it for several hours or overnight before streaking suspensions on isolation media. Several isolations should be attempted from each specimen taken from a disease outbreak.

Successful isolation depends largely in selecting that part of the lesion in which the pathogen is most likely to be present, and in which contaminating or secondary pathogenic micro-organisms are likely to be absent or few in number. It is also dependent upon working with material which is fresh and free from adhering soil particles. If material free from soil is unobtainable, wash the lesion in a stream of tap water. For deep-seated infections or woody tissue, surface disinfection by wiping with 70% ethanol may reduce potential contaminants. This is not usually suitable for spots on delicate herbaceous leaves since the ethanol may be absorbed and the pathogen killed. Further details of isolation methods are discussed under specific diseases.

Plates of isolation media should be dried after pouring, either by inverting and propping the base on the edge of the inverted lid for 24 hours in a clean incubator or by removing the lids in a laminar flow cabinet for 30 minutes. This frees them of surface moisture which can allow motile bacteria to spread over the plate making discrete, well-separated colonies difficult to obtain.

1.4.1 Inoculation of isolation media

A loopful of the suspension of plant tissue in sterile water (1.3) should be streaked on the surface of a plate of the appropriate agar medium. This is done by rubbing a nichrome or platinum loop of 24 s.w.g. (made by forming a loop round a glass rod and then twisting the free end round the standing end as shown in Fig. 1.1) in a zigzag fashion on a quadrant of the plate starting from the circumference. Further quadrants should be progressively streaked, each overlapping the last at the circumference. The loop should be flamed and then cooled by touching the agar surface between streaks; this assists in the separation of individual colonies. The object is to produce numerous well-separated colonies (Fig. 1.2). Although nichrome loops retain their shape better than platinum ones they become rough more quickly and then less effective. Loops can be purchased ready-made.*

*Obtainable from Medical Wire & Equipment Co. (Bath) Ltd, Potley, Corsham, Wilts, UK.

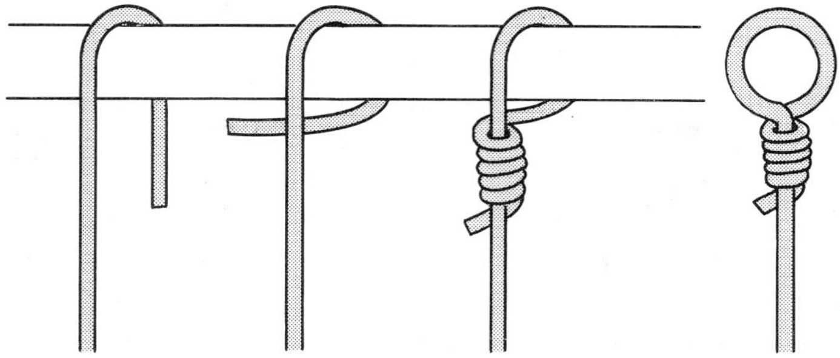

Fig. 1.1. Preparation of wire inoculating loop using 24 swg nichrome or platinum wire by the use of a 4 mm diameter glass rod.

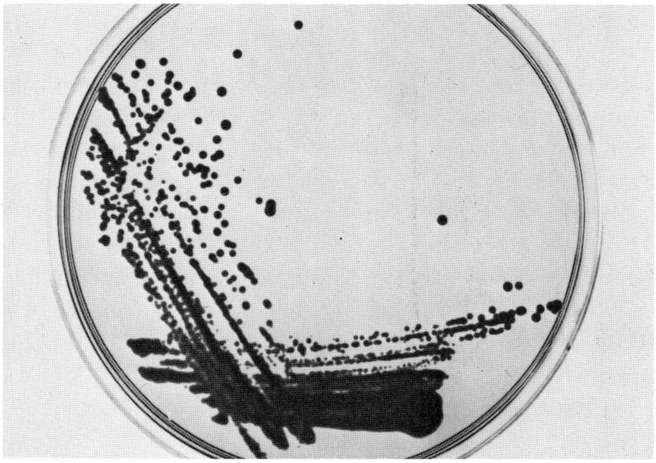

Fig. 1.2. Isolation plate. Growth of isolated colonies of *Clavibacter michiganense* subsp. *michiganense* from infected tomato vascular tissue. The initial 'wash-out' streak of confluent growth may be a mixed culture. Crown copyright. Used by permission of the Controller of HMSO.

Where few or no bacteria are seen in the initial microscopic examination and the symptoms suggests a bacterial disease, some of the diseased tissue should be placed in a separate covered container (e.g. a small Petri dish), comminuted in a drop of sterile water and then streaked after 2–4 hours or even, in some cases, overnight.

For material in which microscopic examination shows very mixed populations of bacteria, the comminution may be suitably diluted in sterile water — two tenfold steps are usually appropriate — and both comminution and the dilution(s) streaked on separate plates.

1.5 Incubation

Almost all plant pathogenic bacteria grow well at 25–27°C, and for most isolations an incubation temperature within this range is adequate. Most coryneform plant pathogens have slightly lower temperature optima than other plant pathogenic bacteria and should, where possible, be incubated at 23–25°C. The optimum growth temperature for *Clavibacter michiganense* subsp. *sepedonicum* is 21°C. *Xanthomonas populi* also has a relatively low optimum growth temperature and should also be incubated at 21°C.

Isolation plates, except pectate gels, should be incubated in the inverted position to prevent condensation of water vapour on the agar surface.

1.6 Interpretation of isolation plates

Apart from symptoms and selection of the specimen for isolation, the isolation plate is the most important stage in many diagnoses. The ideal isolation plate is one with a pure culture of the pathogen with many, well-separated colonies (Fig. 1.2). Sometimes this is not the case, particularly if saprophytic bacteria are present; these usually grow rapidly on non-selective isolation media and can swamp colonies of the pathogen. But if adequate precautions were taken in preparing specimens for isolation (see 1.2, 1.3, 1.4 above) and the specimen was from or of a plant at an early stage of disease development, the pathogen should be the dominant organism on the plate and there should usually be no question as to which bacterium is the pathogen. Most plant pathogenic bacteria, although growing more slowly than saprophytic contaminants, should still produce visible colonies within 3–4 days and often less. However, some species are slow-growing or fastidious; these include all the coryneforms, which usually require at least 4 days incubation. Some, such as *Clavibacter michiganense* subsp. *sepedonicum*, *Erwinia tracheiphila*, *Xanthomonas ampelina* and *X. populi* are particularly fastidious on initial isolation. Colonies may take well over a week to develop, and by that time may be overgrown by saprophytic bacteria, especially from rotted tissue. Often they do not form discrete colonies on initial isolation plates, growing only as fluidal or mucoid growth in the washout streak (Fig. 1.2) or later streaks (Fig. 1.3). When the pathogen is suspected to be fastidious and slow-growing, some bacterial growth from the isolation streaks should be taken, well away from any obvious contaminant bacteria and, as soon as it is seen appearing on the isolation plate, should be dispersed in sterile water to give a faintly turbid suspension and re-streaked to fresh plates of the same medium to improve the chance of isolation.

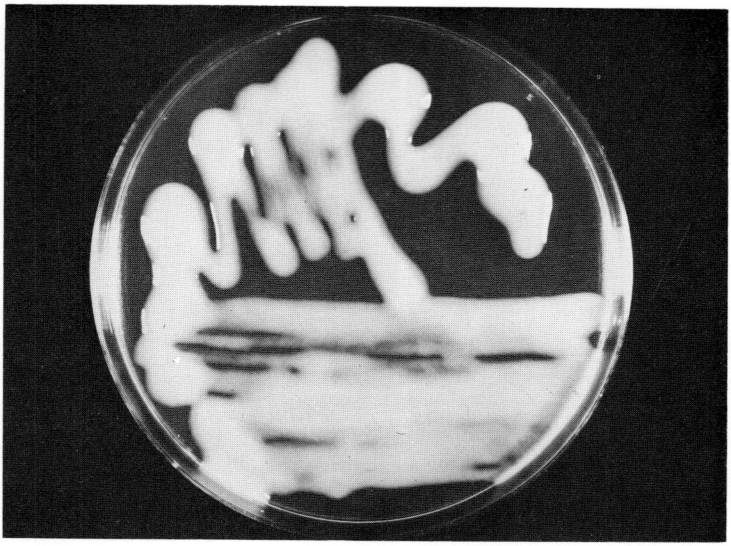

Fig. 1.3. Fluidal growth of *Xanthomonas populi* on an isolation medium (yeast peptone glucose agar). Crown copyright. Used by permission of the Controller of HMSO.

It is perhaps tempting for the inexperienced worker who has isolated a mixture of bacterial colony types on a plate with no obviously dominant colony type to purify a wide selection of these and attempt to identify each in the hope that one is the pathogen. Where yellow-pigmented bacteria, or green-fluorescent pseudomonads are present the temptation is greater. Apart from exceptional cases indicated later in this book, this should not be done. It is better to re-isolate from the original plant material, perhaps on a wider range of media and pay particular attention that the initial microscopic examination is carefully and thorougly followed (1.3).

Examination of isolation plates may indicate that a bacterial pathogen is not involved. Whether or not a report that a bacterial pathogen is not considered to be involved should be made without further examination must be influenced by such factors as the experience of the diagnostician, the suitability of the specimen for isolation and how many bacterial colonies were present on the isolation plate. If the specimen is in poor condition and no more material is available it may be more prudent to report 'no bacterial pathogen isolated'. The same report may be advisable if an inexperienced diagnostician suspects a bacterial disease from symptoms but can not isolate bacteria which warrant further investigation. Again, if many bacteria are present but repeated attempts to isolate a pathogen fail the diagnostician may prefer to give a cautious report. He should alert his client to the possibility of a bacterial disease which needs further field investigation.

A preliminary diagnosis of a particular pathogen is often possible at an early stage, based on symptomatology and the dominant presence of organisms with characteristic colonies on the isolation plates; in some cases such a diagnosis will be sufficient and no further work will be needed on the specimen. If further work is needed, isolated colonies of bacteria to be investigated should be well-suspended in sterile water (preferably using a rotary mixer) and re-streaked onto a suitable medium to ensure purity. This is particularly important where selective media have been used; isolated colonies of apparently one bacterium may contain other bacteria which are inhibited but are still viable. It must also be remembered that colony variation may occur. Such variation may be stable or unstable. King's medium B in particular appears to stimulate such variation and plates may appear to have more than one bacterium on them when in fact they have merely more than one colony form of the same bacterium.

1.7 Diagnostic tests

Many diagnostic and identification tests are based upon the nutritional and physiological properties of bacteria. Some of these such as Gram reaction, ability to use oxygen as a terminal electron acceptor, presence of cytochrome-c oxidase, accumulation of poly-β-hydroxybutyrate, arginine dihydrolase activity and pectate lyase activity are based upon important taxonomic features and usually have great reliability. Generally speaking, individuals of a given species should react similarly when tested for such characteristics.

Conversely, the ability to utilize or produce acid from particular sugars, organic acids or amino acids often varies between isolates of the same species or depends upon the basal medium used and particularly on the length of incubation time. For such tests this needs to be taken into account by looking at the results of tests on a range of such characters and matching the pattern obtained to the pattern described in such texts as Bergey's Manual of Determinative Bacteriology (Buchanan & Gibbons, 1974; Krieg & Holt, 1984) for a range of bacteria; that is, one looks for *overall* similarity of such characters and disregards divergence by individual isolates in individual characters. This process is very time-consuming and laborious and usually unsuitable for use by diagnostic laboratories. Relatively little value has been given to it in this text except where more reliable characteristics are unavailable.

Too much emphasis should not be given to any one test or type of test, particularly by inexperienced workers. The value of the various types of test differs with the organism and an attempt has been made to take account of this in the diagnostic sections (Chapter 3).

Apart from nutritional and physiological tests, a wide variety of other testing procedures may be used. Each has its advantages and its disadvantages and has particular relevance to certain groups of bacteria.

The use of bacteriophages can be a useful diagnostic aid for some groups of pathogens but they are perhaps of greater use in typing specific, known isolates in epidemiological studies. If available, however, there is no reason why their use should not be used in conjunction with other tests. One report, as yet unconfirmed generally, indicates their usefulness for diagnosis of *Pseudomonas syringae* pv. *tomato* (Cuppels, 1984).

Serology has become an important tool in the diagnosis of many pathogens. However, an appreciation of the chemical basis for antibody specificity and the methods by which this can be demonstrated is a prerequisite for the useful application of serological techniques to diagnosis. Its major limitation is the fact that the outer surface of bacterial cells is antigenically heterogenous — unlike virus particles with their repetitive sub-units. Hence an antiserum produced against whole bacterial cells will contain many different cell-wall and flagella antibodies as well as antibodies to any endogenous antigens released during immunization. Some of the antigens of one pathogen are frequently common to other related or even unrelated bacteria, and cross reactions will be then obtained. Antisera can often be made more specific, for example either by cross-absorbing with heterologous antigen or by dilution in the hope of diluting out cross-reacting antibodies. Hopefully, the use of monoclonal antibodies will improve the performance of serological diagnosis by enabling antisera to individual antigens to be prepared which can then be combined to give a pooled antiserum of the required specificity. This is a very laborious process and requires extensive knowledge of the antigenic make-up of pathogens and their antigenically related saprophytes.

1.8 Host specificity and pathogenicity

Most plant pathogenic bacteria are host specific or attack only a few plant species. The coryneforms, most non-pectolytic erwinias, the green-fluorescent pseudomonads and xanthomonads have narrow host ranges. But not all plant-pathogenic bacteria are host specific. Several have a wide host range. These include the agrobacteria, pectolytic erwinias, *Rhodococcus fascians, Pseudomonas solanacearum, P. avenae, P. andropogonis, P. syringae* pv. *syringae, P. viridiflava, P. marginalis, P. cichorii,* and *P. fluorescens.* Other bacteria have as hosts several closely related species in the same family, e.g. *Erwinia amylovora* and members of the *Pomoideae* in the family *Rosaceae, Xanthomonas campestris* pv. *graminis* and members of the family *Gramineae.*

Some bacterial species are unspecialized, opportunist pathogens;

where plants are grown, or plant products stored, under adverse conditions such bacteria may infect them and cause disease. Many are pectolytic bacteria normally found in soil and on root or leaf surfaces — *Pseudomonas marginalis*, *Pseudomonas viridiflava* and pectolytic *Pseudomonas fluorescens* are examples. Other pectolytic bacteria not easily proven to be pathogenic include *Bacillus* spp. (particularly *B. polymyxa*) and *Cytophaga johnsonae*, often referred to as *Flavobacterium pectinovorum* — a species not in the Approved Lists (Skerman *et al.*, 1980). Under certain conditions many soil bacteria can act as minor pathogens of roots, causing, *in toto*, substantial yield losses. *Erwinia herbicola*, a very common epiphyte on plant surfaces, is also occasionally known to be pathogenic.

Within these constraints, a positive host test is often the most certain determinate character and is usually required for confirmatory diagnosis. When carried out rapidly on readily available material, host tests are extremely valuable. Some experience is necessary in selecting the host organs (fruits, leaves, stems, whole organs, tissue slices, etc.) to be used, in determining the most suitable concentration of cells for the inoculum, in the choice of incubation conditions and in the interpretation of results. Negative reactions should always be treated cautiously. Similarly, weak, atypical symptoms may not reflect a natural host reaction.

1.9 Diagnostic strategy

Chapters 2 and 3 provide a diagnostic guide for most diseases caused by bacteria, that is based upon symptoms, isolation and practical identification schemes using relatively few but important key tests. It is suggested that the diagnostician should, after having determined by microscopic examination that the disease affecting the specimen(s) he/she is investigating is likely to have a bacterial cause (see 1.3), refer first to section 2.7 to ascertain which pathogens are the most likely causal agents. Referring to the appropriate sections in Chapter 3 will then normally reduce the list of candidate pathogens to one or two, which can then be further investigated. Rarely is there a need, nor would it be wise, to make isolations from material without having a hypothetical cause in mind.

1.10 Preservation of cultures

Any diagnostic laboratory needs a collection of cultures of known identity for use as positive controls in the confirmatory tests used to check presumptive diagnosis; this is particularly true of pathogenicity tests. Cultures cannot be maintained by storage on media at room temperature without frequent subculturing and without the probability that cultures will lose pathogenicity or become atypical in other respects. There are many methods for preserving cultures with little risk of such changes. These include storage:

1 in sterile distilled water at 10°C or at room temperature (de Vay & Schnathorst, 1963);
2 on agar slopes under oil;
3 in sterile soil;
4 on agar slopes at −20°C (Stead, 1974);
5 on sterile glass beads in vials stored at −70°C (Feltham *et al.*, 1978);
6 as lyophilized cultures (Lelliott, 1965).

There is little or no published evidence for the effectiveness of some of these methods with a wide range of pathogens, and the only extensive published comparison is between lyophilization, storage under oil and in soil (Lelliott, 1965). Probably the best method for general use is lyophilization.

Most plant-pathogenic bacteria remain viable in the lyophilized state for long periods of time, usually in excess of 10 years and often for more than 50 years. Where lyophilization equipment is not available, preservation either in soil or under oil is recommended. In the case of specific pathogens for which there is no evidence for the effectiveness of any methods, storage by at least two methods is advisable. Table 1.1 indicates which of these methods is preferred for each major group of bacterial pathogens.

1.10.1 Preservation under oil

Streak the surface of agar slants in bijou bottles heavily with the bacterium and incubate at 25°C until well grown. The cultures should then be covered with sterile liquid paraffin oil which has been sterilized at 121°C for 15 minutes and dried overnight in an oven at 100°C. The level of the oil should be well above the top of the agar. These cultures should be stored at 4°C but will remain viable for a long time even at ambient temperatures. When subculturing, scrape off some of the growth with a small loop and withdraw the loop carefully through the oil layer to ensure that the growth is not detached from the loop as it is withdrawn from the oil. This occurs particularly commonly with xanthomonads, and for these it is advisable to remove a small part of the agar with the culture growth.

1.10.2 Preservation in soil

Cultures in soil should be prepared by pipetting approximately 0.5 ml of a heavy, nutrient-broth suspension from a well-grown agar slope culture on to about 3 g of sterile garden soil in small, plugged test tubes (75 × 10

Table 1.1. Methods of preservation of plant pathogenic bacteria other than by lyophilization.

Bacterium	Preferred method of preservation	Growth medium	Subculturing interval (years)
Agrobacterium spp.	Soil/oil	NA	1–2
Coryneforms	Soil/oil	YDCA	1
Erwinia spp. (other than soft rot group)*	Oil/soil	NA	2
Erwinia spp. (soft rot group)†	Oil slopes at −20°C/ distilled water	NA	1
Pseudomonas spp. (green fluorescent types)	Soil/oil	NA	1
Pseudomonas spp. (non-fluorescent types)	Oil	YDCA	0.5
Xanthomonas spp.††	Oil/soil	YDCA	1

NA = Nutrient agar
YDCA = Yeastrel dextrose chalk agar
**Erwinia tracheiphila* rapidly loses virulence during storage.
†*Erwinia chrysanthemi* is less viable than *E. carotovora*.
††*Xanthomonas ampelina* and *Xanthomonas populi* do not remain viable for more than a few months.

mm) or bijou bottles. The soil should have been air-dried, put through a 2 mm mesh sieve and autoclaved at 128°C for 1 hour on 3 successive days. Store the cultures at 4°C and recover from them by sprinkling a few crumbs of soil from a sterile loop or spatula onto a dried agar plate.

1.10.3 Maintenance of cultures by lyophilization

Lyophilization represents perhaps the best method of culture maintenance for most bacterial plant pathogens and is preferred by most collections of bacteria including the National Collection of Plant Pathogenic Bacteria (NCPPB).

Although most plant-pathogenic bacteria can be stored for many years, it is useful to have an estimate of the half-life for species and pathovars, i.e. the time taken for the viable population to be halved. For most plant pathogens this is between 9 months and 3 years. If the viable population immediately after lyophilization is also estimated, the shelf-life of lyophilized cultures can be determined. Lelliott (1965) determined the half-life of many lyophilized pathogens stored at room temperature. Storage at lower temperatures increases the half-life, and work in the NCPPB using the same method but with storage at 4°C has given the average expected half-lives and safe shelf lives from a population of 10^9/viable cells/ampoule immediately after lyophilization, shown in Table 1.2; this assumes that viability should not fall to less than 10^3 viable cells/ampoule.

Table 1.2. Viability of lyophilized cultures of plant-pathogenic bacteria.

	Half-life (years)	Shelf-life (years)
Coryneforms	3.0	>50
Agrobacterium	0.8	16
Erwinia spp.	0.9	18
E. chrysanthemi	0.8	16
E. tracheiphila	0.8	16
Pseudomonas spp.	0.9	18
P. solanacearum	0.8	16
P. rubrisubalbicans	0.8	16
Xanthomonas campestris	2.0	40
X. albilineans	0.8	16
X. populi, *X. ampelina*, *X. axonopodis*	Not determined but low	

1.10.3.1 **Lyophilization procedure**

Streak the culture on to suitable media and check for purity and authenticity. Reduce the possibility of selecting a mutant by transferring several, well-isolated colonies to one slope of the selected medium. When grown, suspend in sterile water and transfer the growth to a large, flat Roux bottle of the same agar medium. Incubate until cells are approximately in the stationary phase of growth (2–3 days for rapidly-growing species). Wash off the growth in 20 ml of sterile lyophilization medium containing 7% sucrose, 7% peptone (sterilized as separate 14% solutions and then mixed in equal parts) to which have been added sterile glass beads to aid the removal of bacterial growth from the agar. Decant from the beads and centrifuge the dense suspension (10 000 g for 10 minutes). Decant the supernatent and resuspend the bacteria in 3–4 ml of the sucrose peptone medium.

Check glass ampoules for cracks, push a small pledget of cotton wool to the bottom of each ampoule followed by a printed label made of filter paper and plug loosely with cotton wool. This may be colour coded to denote the batch number. Sterilize by autoclaving at 121°C for 15 minutes.

Introduce aseptically 0.02 ml of the bacterial suspension to each ampoule and replace the cotton wool plug. Trim the plugs by cutting them off flush with the top of the ampoule and push down to standard depth approximately 3 cm from the base.

Constrict the tubes just above the plugs. Place the ampoule tubes on a secondary drying head attached to a freeze-drying apparatus.* Evacuate to more than 1.33 Pa and dry in the dark for approximately 18–24 hours.

Seal the ampoules *in vacuo* at the constriction by flaming with a gas-fired sealing torch. Allow to cool. Store in the dark at 4°C.

Other methods of lyophilization use two stages: a primary stage under centrifugation or from the frozen state and a secondary stage after constriction. The data given in Table 1.2 refer only to single stage evaporative freeze-drying.

1.10.3.2 **Purity**

Crack an ampoule from a lyophylized batch, add sterile 0.1% peptone solution to reconstitute the bacteria and pour the entire contents on to an agar plate of a suitable growth medium. Spread the culture to obtain a large surface area of primary washout (Fig. 1.2) and well-separated dilution streaks. Incubate at 27°C for 10 days. Observe for colony variation and contamination.

1.10.3.3 **Viability counts immediately after lyophilization**

Crack an ampoule from a lyophylized batch, add sterile 0.1% peptone solution to reconstitute the bacteria and pour the entire contents into 20 ml of sterile 0.1% peptone solution. Mix well and dilute 0.1 ml of this to another 20 ml of sterile peptone solution. Transfer 0.1 ml of this to 20 ml of a suitable clear, sterile molten medium at 45°C (not yeast dextrose chalk agar medium). Mix well, pour immediately into a Petri dish, allow to set, incubate at 27°C for 2–3 days. Count visible colonies and determine the viable population per ampoule. A series of masks to give viewing windows of 1 mm^2, 10 mm^2 and 100 mm^2 can be used to sample areas of the plate for counting. Window size and number of areas should be adjusted to count at least 300 colonies.

The dilution stage(s) can be omitted as appropriate if counts are made after storage and the expected number of cells/ampoule is low.

1.10.3.4 **Authenticity**

Streak from the first peptone solution dilution of the viability count (1.10.3.2 above) on to several suitable media to obtain well-separated

*Obtainable from Edwards High Vacuum Ltd, Crawley, Sussex, UK.

colonies. Incubate at 27°C for 3 days and compare the morphology and other characteristics of colonies with those recorded for the culture before lyophilization. Perform other determinative tests appropriate for the bacterium lyophilized.

1.11 Laboratory safety

All bacteria growing on microbiological media should be regarded as being potentially pathogenic to humans, however unlikely this is, and should be handled appropriately. Several plant pathogenic bacteria are known to be associated also with human disease.

The essential feature of laboratory safety is good practice. As well as good practical techniques this also includes knowledge of the mechanisms through which infection may occur and an awareness of the potential risks for each organism handled. Good practice also implies that the laboratory has adequate equipment.

There are now many excellent references on safety in the microbiology laboratory and all workers should be familiar with their recommendations. These include: General laboratory safety — Chatigny (1961), Collins *et al.* (1974), Department of Health and Social Security (1972), Emmett Barkley (1981), Melling and Allner (1981), Laboratory operations — Grieff (1969), Sykes (1967).

This section reviews those plant pathogenic bacteria which have also been implicated with human disease and covers the most important operations which may place the operator at risk.

Bacteria implicated with both plant and human disease include *Agrobacterium tumefaciens*, *Erwinia herbicola* (syn. *Enterobacter agglomerans*), *Pseudomonas aeruginosa*, *Pseudomonas cepacia* and perhaps some *Bacillus* spp. Most of these are opportunistic human pathogens or have been occasionally isolated from diseased human tissues as with *Agrobacterium tumefaciens* in post-surgical operation infections. *Pseudomonas aeruginosa*, however, is a particularly dangerous human pathogen also reported to be an opportunistic plant pathogen.

The following laboratory safety practices should be routinely used in all bacteriology laboratories:

1 Do not eat, drink or smoke in, or allow food and drink into the laboratory.
2 Wash hands frequently and always before leaving the laboratory.
3 Keep laboratory and incubator doors closed.
4 Disinfect benches daily and always immediately after culture spillages.
5 Minimize production of aerosols containing bacteria, and where this is unavoidable, ensure adequate containment.
6 Autoclave or otherwise disinfect all contaminated items as soon as possible before washing or re-using them. Do not leave cultures unnecessarily on the bench.

7 Do not pipette by mouth; use safe mechanical pipettes.
8 Disinfect and autoclave all hypodermic syringes and needles before disposal.
9 Always wear suitable protective clothing in the laboratory, e.g. clean laboratory coat; remove before leaving the laboratory.

There are some laboratory procedures which are commonly used in plant bacteriology laboratories which may present risks to the user unless suitable precautions are taken. Bacterial aerosols for spraying plants should be contained preferably in a biological safety cabinet. These operations should not be performed in horizontal laminar airflow cabinets which exhaust directly towards the operator. In instances where the use of biological safety cabinets is impractical, e.g. in field experiments, operators should wear autoclavable protective clothing and a face mask and goggles.

Many of these safety techniques will also obviate contamination of cultures.

Physical methods for disinfection are the most efficient, e.g. autoclaving at 121°C for 15 minutes. When these methods are impractical, chemical disinfection can be achieved with any one of a range of chemicals including the following:
1 Phenolic compounds are most used, but some are skin irritants and have an unpleasant smell. Synthetic phenols (such as Hycolin) without these drawbacks may be preferred. They are usually used at concentrations of 1–5% in water for up to 10 minutes.
2 Ethyl alcohol — usually used at dilutions of 70–85% in water for up to 10 minutes.
3 Quaternary ammonium compounds — usually used at dilutions of 0.1–2% in water for up to 10 minutes. They are non-corrosive but are inactivated by organic matter.
4 Chlorine compounds — most household bleaches contain 5–10% available chlorine. A 1:100 dilution of this in water makes a useful disinfectant (500–1000 ppm available chlorine) but chlorine compounds are corrosive and are inactivated by organic matter.

2

Symptoms and their Expression

2.1 Introduction

Symptomology is the starting point for diagnosis. In fact, a narrow dictionary definition of diagnosis is 'identification of disease by means of ... symptoms'. The diagnostician needs to be familiar with the full range of symptoms (including those that are atypical) for the disease he/she commonly sees, and this can only be gained by practice and by checking visual diagnoses by subsequent isolation and identification of the pathogen. The descriptions of symptoms presented here indicate only the major features of diseases and are no substitute for self-training in the field and laboratory. More detailed descriptions may be obtained from Smith (1905, 1911, 1914, 1920), Elliott (1951), Dowson (1957), Stapp (1961), Fahy and Persley (1983) and the various Commonwealth Mycological Institute Descriptions of Pathogenic Fungi and Bacteria and by consulting the original descriptive papers.

Section 2.7 of this chapter lists the genera of hosts affected by bacterial diseases and the pathogens associated with particular symptoms in each genus. By reference to this section, and then to the more detailed descriptions of symptoms in Chapter 3, it should be possible to restrict the possible cause of any bacterial disease to a few pathogens which can then be investigated. It should be realized that the key in section 2.7 is not comprehensive. The host range of many bacterial pathogens and the range of symptoms they cause are incompletely known; this applies particularly to diseases that occur only in the tropics or sub-tropics.

2.2 Hyperplasias

Hyperplasias are diseases in which abnormal growth has occurred and include fasciations, root and shoot proliferation, galls and hyperplastic cankers. All appear to arise as a result of impaired hormonal or growth regulatory system; in the case most studied — crown gall caused by *Agrobacterium tumefaciens* — they result from the incorporation of genes borne on the plasmids of the bacterium into the host genome.

Galls are tumorigenic swellings and may be found on stems, leaves and roots sometimes associated with proliferation of shoots and/or roots. Their surfaces are usually cracked, sculptured or textured. They are extremely variable in size; crown gall may be found as pin head galls on leaves or as large stem or root tumours weighing several kilograms. Fasciation comprises shoot or root proliferation often with flattening and fusion. Plants often survive infection by gall-inciting bacteria, although growth can be seriously affected.

Most bacteria that can cause hyperplasias have a wide host range.

2.3 Leaf, stem and fruit spots and necroses

These are the most common symptoms produced by bacteria. When plant pathogenic bacteria enter susceptible stems, leaves and fruits, through stomata, lenticels, hydathodes, trichomes, or wounds they can multiply and spread in the intercellular spaces. Colonized tissue usually dies fairly rapidly and becomes necrotic. Depending on the pathogen involved and often the physiologic state of the host the spread of the pathogen in tissue is either soon stopped, forming spots, or continues to form large, necrotic areas (necroses). If the spot is limited by veins (as it often is on leaves) the spot is angular in shape. Many bacteria that produce leaf spots, particularly those in Pseudomonas group Ia, also produce toxins which diffuse into the surrounding healthy tissue and result in the information of chlorotic pale-green or yellow haloes around the necrotic spots. These haloes usually have very few, if any, viable cells of the pathogen in the chlorotic tissue. Opportunist pathogens such as *Pseudomonas cichorii, P. viridiflava, P. marginalis* and *Erwinia carotovora* and *E. chrysanthemi* produce more spreading lesions. Localized infections can coalesce and the resultant disease is often referred to as a blight. In young shoots, infection of the shoot tip or via a blossom truss usually results, under favourable conditions, in rapid colonization of parenchyma and collenchyma tissues. The tip then withers and dies as the infection progresses down the shoot. This symptom is referred to as dieback.

2.4 Cankers

Bacteria may gain access to parenchyma tissues of stems beneath the bark or epidermis via wounds, lenticels, or leaf scars. Multiplication in these tissues is usually accompanied by production of extracellular polysaccharides. Infected tissue dies and colonization of adjacent healthy tissues, which is sometimes relatively slow, is governed by climatic conditions. When spread of the bacteria through tissues ceases and the disease becomes quiescent, cankers are often clearly delimited from the healthy tissue and infected tissues often appear sunken and discoloured. Bacteria do not readily spread to underlying vascular tissues and so cankers tend to be shallow. Bacterial multiplication can increase again when environmental conditions become favourable and for many diseases of deciduous trees cankers provide a primary source of inoculum for colonization of newly-produced shoots, leaves and blossom. Rapid multiplication of cells is often accompanied by production of copious quantities of extracellular polysaccharides loosely termed 'ooze' which exudes from infected areas and is transmitted to other tissues by rain splash and/or insects. The bacterial exudates are a common feature of many diseases other than hyperplasias and can occur on leaf and fruit lesions.

2.5 Systemic diseases

Various internal tissues may be infected and further spread of the pathogen may continue within these tissues. The majority of systemic diseases involve transmission of bacteria in the xylem which usually results in wilting due either to the blocking of vessels by production of extracellular bacterial polysaccharides, or to the production of toxins that interfere with the permeability of the cell membranes of the leaf tissue supplied by the infected xylem. Vascular infections do not always result in wilting although plant vigour is usually impaired resulting in stunting, chlorosis and consequent yield loss. Some soft rot diseases have a systemic phase in which most stem tissues are infected but remain fairly intact. In some diseases spread may be so rapid in collenchyma and/or parenchyma tissues that symptoms similar to those of systemic diseases may be present, e.g. tomato pith necrosis caused by *Pseudomonas corrugata*.

2.6 Soft rots

Many plant pathogenic bacteria produce one or more pectic enzymes, which are usually polygalacturonases or pectate lyases. These enzymes degrade pectic substances in the middle lamella and cell walls. This activity results in leakage from, and death of cells and, with certain pathogens, in tissue-maceration commonly known as soft rot. The rates at which these enzymes are produced and the efficacy with which they can degrade pectic substances varies with the type of bacterium producing them. Some *Erwinia* species are particularly effective at rotting such storage tissues as potato tubers and turnip roots. Soft-rotting bacteria gain access to succulent tissues through lenticels, stomata and wounds. Rapid colonization of infected tissues usually occurs in conditions unfavourable to the host (e.g. wounding, high temperatures, high turgidity, anaerobic conditions). Storage organs (bulbs, tubers, etc.) and succulent leaves are particularly susceptible to soft rotting.

Although many plant pathogenic bacteria can produce pectolytic enzymes, the major soft-rotting bacteria belong to certain *Erwinia* and *Pseudomonas* species, e.g. *E. carotovora, E. chrysanthemi, P. viridiflava* and *P. marginalis*.

2.7 Key to pathogens

Host genus	Symptom	Possible pathogen(s)
Aechinea	Rotting of leaves and stems	*Erwinia chrysanthemi*
Agloanema	Rotting of leaves and stems	*E. chrysanthemi*
Agropyron	Leaf striping	*Pseudomonas avenae*
	Leaf spotting	*P. avenae*
Agaricus	Blotching of caps	*P. gingeri*
		P. tolaasii
	Oozing from gills	*P. agarici*
	Pitting of caps	*Bacillus polymyxa*
Allium	Soft skin, rotting	*Pseudomonas cepacia*
	Soft rotting	*P. gladioli* pv. *alliicola*
		Erwinia carotovora subsp. *carotovora*
		E. herbicola
	Leaf necrosis	*Pseudomonas syringae* pv. *porri*
		P. syringae pv. *syringae*
Ananas	Fruit rotting	*Erwinia ananas*
		E. chrysanthemi
	Pink discoloration	*Gluconobacter oxydans*
Anthurium	Watersoaking, leaf and stem necrosis	*Xanthomonas campestris* pv. *dieffenbachiae*
Antirrhinum	Leaf spotting and necrosis	*Pseudomonas syringae* pv. *antirrhini*
Apium	Leaf spotting and necrosis	*P. syringae* pv. *apii*
		P. cichorii
		P. marginalis
Arachis	Wilting	*P. solanacearum*

Host genus	Symptom	Possible pathogen(s)
Areca	Leaf spotting and/or striping	*Xanthomonas campestris* pv. *vasculorum*
Avena	Leaf and sheath spotting	*Pseudomonas syringae* pv. *coronafaciens*
		P. s. pv. *striafaciens*
	Glume spotting	*P. s.* pv. *coronafaciens*
	Leaf striping	*P. s.* pv. *striafaciens*
	Leaf death	*P. s.* pv. *striafaciens*
Axonopus	Leaf striping	*Xanthomonas axonopodis*
	Leaf drop	*X. axonopodis*
	Stunting and withering	*X. axonopodis*
Begonia	Stunting and wilting	*Erwinia chrysanthemi*
		Xanthomonas campestris pv. *begoniae*
	Leaf spotting and necrosis	*X. c.* pv. *begoniae*
		Pseudomonas cichorii
	Soft rotting of leaves and stems	*Xanthomonas campestris* pv. *begoniae*
Berberis	Leaf spotting	*Pseudomonas syringae* pv. *berberidis*
	Shoot dieback	*P. s.* pv. *berberidis*
Beta	Rounded galling usually at soil level on root	*Agrobacterium tumefaciens*
	Silvering and wilting of leaves of red beet, usually of seed plants	*Curtobacterium flaccumfaciens* pv. *betae*
	Rotting of roots	*Erwinia carotovora*
		E. chrysanthemi
		Pseudomonas marginalis
Boerhavia	Leaf necrosis	*Xanthomonas campestris* pv. *campestris*
	Veinal necrosis	*X. c.* pv. *campestris*
	Leaf yellowing	*X. c.* pv. *campestris*
Bougainvillea	Leaf spotting	*Pseudomonas andropogonis*
	Leaf striping	*P. andropogonis*
Brachutica	Stunting, yellowing, wilting	*Erwinia chrysanthemi*
Brassica	Leaf spotting	*Pseudomonas syringae* pv. *maculicola*
		P. viridiflava
	Leaf marginal necrosis	*Xanthomonas campestris* pv. *campestris*
		X. c. pv. *armoraciae*

Host genus	Symptom	Possible pathogen(s)
	Leaf yellowing	*X. c.* pv. *campestris*
	Veinal necrosis	*X. c.* pv. *campestris*
		X. c. pv. *armoraciae*
	Soft rotting of stems and roots	*Erwinia carotovora*
		Pseudomonas marginalis
		P. viridiflava
Bromus	Leaf striping	*P. avenae*
	Leaf spotting	*P. avenae*
		Xanthomonas campestris pv. *cerealis*
Camellia (C. japonica)	Leaf spotting	*Pseudomonas syringae* pv. *theae*
	Shoot dieback	*P. s.* pv. *theae*
Capsicum	Leaf spotting	*Xanthomonas campestris* pv. *vesicatoria*
	Fruit spotting	*Clavibacter michiganense* subsp. *michiganense*
		Xanthomonas campestris pv. *vesicatoria*
	Fruit rot	*Erwinia carotovora*
	Wilting	*Pseudomonas solanacearum*
		Clavibacter michiganense subsp. *michiganense*
Chrysanthemum	Stunting and wilting	*Erwinia chrysanthemi*
	Leaf and stem necrosis	*E. chrysanthemi*
		Pseudomonas syringae pv. *syringae*
		P. cichorii
		P. marginalis
		P. viridiflava
	Galling of leaves and stems	*Agrobacterium tumefaciens*
	Fasciation	*Rhodococcus fascians*
Cichorium	Leaf necrosis	*Pseudomonas cichorii*
Citrus	Leaf spotting	*P. syringae* pv. *syringae*
		Xanthomonas campestris pv. *citri*
	Branch cankering	*Pseudomonas syringae* pv. *syringae*
		Xanthomonas campestris pv. *citri*
Coffea	Leaf spotting	*Pseudomonas syringae* pv. *garcae*

Host genus	Symptom	Possible pathogen(s)
	Shoot dieback	*P. s.* pv. *garcae*
Colocasia	Wilting and rotting	*Erwinia chrysanthemi*
Corylus	Leaf spotting	*Xanthomonas campestris* pv. *corylina*
	Fruit spotting	*X. c.* pv. *corylina*
	Shoot dieback	*X. c.* pv. *corylina*
	Bud necrosis	*X. c.* pv. *corylina*
Cotoneaster	Blossom blight, leaf and shoot necrosis	*Erwinia amylovora*
Crataegus	Blossom blight, leaf and shoot necrosis, cankering	*E. amylovora*
Cucumis	Wilting	*E. tracheiphila*
	Leaf and stem spotting	*Pseudomonas syringae* pv. *lachrymans*
		Xanthomonas campestris pv. *cucurbitae*
	Fruit spotting	*Pseudomonas syringae* pv. *lachrymans*
Cucurbita	Leaf spotting	*Xanthomonas campestris* pv. *cucurbitae*
		Pseudomonas syringae pv. *lachrymans*
Cyamopsis	Leaf and stem necrosis	*P. s.* pv. *syringae*
Cypripedium	Leaf spotting	*Erwinia cypripedii*
	Leaf and stem necrosis	*E. cypripedii*
Dactylis	Yellow slime on upper parts	*Clavibacter rathayi*
Dahlia	Stunting and wilting	*Erwinia chrysanthemi*
	Leaf necrosis	*E. chrysanthemi*
		Pseudomonas cichorii
		P. viridiflava
	Fasciation	*Rhodococcus fascians*
Delphinium	Black leaf spotting	*Pseudomonas syringae* pv. *delphinii*
		P. viridiflava
Dianthus	Stunting and wilting	*Erwinia chrysanthemi*
		Pseudomonas caryophylli
	Stem cracking	*Erwinia chrysanthemi*
		Pseudomonas caryophylli
	Leaf spotting	*P. andropogonis*
Dictyosperma	Leaf spotting and striping	*Xanthomonas campestris* pv. *vasculorum*

Host genus	Symptom	Possible pathogen(s)
Dieffenbachia	Soft rotting	*Erwinia chrysanthemi*
	Leaf and stem necrosis	*Xanthomonas campestris* pv. *dieffenbachiae*
Echinocloa	Leaf streaking	*X. c.* pv. *translucens*
Eriobotrya	Stem canker and bud necrosis	*Pseudomonas syringae* pv. *eriobotryae*
Euchlaena	Leaf and sheath spotting	*P. avenae*
		P. andropogonis
	Leaf striping	*P. avenae*
		P. andropogonis
	Stalk rotting	*P. avenae*
Euphorbia	Stunting and wilting	*Erwinia chrysanthemi*
	Watersoaking, defoliation and leaf spotting	*Curtobacterium flaccumfaciens* pv. *poinsettiae*
	Leaf spotting	*Xanthomonas campestris* pv. *poinsettiicola*
Fragaria	Leaf spotting and necrosis	*Xanthomonas fragariae*
	Wilting	*X. fragariae*
	Leaf galling	*Rhodococcus fascians*
Fraxinus	Stem cankering	*Pseudomonas syringae* subsp. *savastanoi*
Freesia	Corm scabbing, leaf necrosis	*P. gladioli* pv. *gladioli*
Geranium	Leaf spotting	*Xanthomonas campestris* pv. *pelargonii*
Gladiolus	Corm scabbing, leaf spotting	*Pseudomonas gladioli* pv. *gladioli*
Glycine	Leaf spotting	*P. syringae* pv. *glycinea*
		P. s. pv. *phaseolicola*
		P. s. pv. *tabaci*
		Xanthomonas campestris pv. *glycines*
Gossypium	Leaf spotting	*X. c.* pv. *malvacearum*
	Stem spotting/rotting	*X. c.* pv. *malvacearum*
	Ball spotting/rotting	*X. c.* pv. *malvacearum*
Hedera	Leaf spotting	*X. c.* pv. *hederae*
Helianthus	Leaf spotting	*Pseudomonas syringae* pv. *helianthi*
Hibiscus	Leaf and stem necrosis	*P. s.* pv. *syringae*
Hordeum	Leaf spotting	*P. avenae*

Host genus	Symptom	Possible pathogen(s)
		P. syringae pv. *atrofaciens*
		Xanthomonas campestris pv. *hordei*
	Leaf striping	*Pseudomonas avenae*
	Glume blotch and blackening	*P. syringae* pv. *atrofaciens*
		Xanthomonas campestris pv. *hordei*
	Seed blackening	*Pseudomonas syringae* pv. *atrofaciens*
Humulus	Rounded galling on stem, usually at soil level or on root	*Agrobacterium tumefaciens*
Hyacinthus	Bulb rotting	*Xanthomonas campestris* pv. *hyacinthi*
		Erwinia rhapontici
	Leaf necrosis	*Xanthomonas campestris* pv. *hyacinthi*
Juglans	Stem cankering	*Erwinia nigrifluens*
		E. rubrifaciens
	Exudation	*E. nigrifluens*
		E. rubrifaciens
	Leaf and shoot spotting and necrosis	*Pseudomonas syringae* pv. *syringae*
		Xanthomonas campestris pv. *juglandis*
	Fruit spotting	*Pseudomonas syringae* pv. *syringae*
		Xanthomonas campestris pv. *juglandis*
Lablab	Leaf, stem and/or pod spotting	*Pseudomonas syringae* pv. *pisi*
Lactuca	Leaf spotting and necrosis	*P. viridiflava*
		P. cichorii
		P. marginalis
		Xanthomonas campestris pv. *vitians*
	Vascular necrosis	*Pseudomonas viridiflava*
		P. marginalis
	Butt rotting	*P. viridiflava*
		P. marginalis
	Soft rotting	*Erwinia carotovora*
		Pseudomonas marginalis
		P. viridiflava

Host genus	Symptom	Possible pathogen(s)
Lathyrus	Leaf, stem and/or pod spotting	*P. syringae* pv. *pisi*
Leersia	Leaf striping	*Xanthomonas campestris* pv. *oryzae*
	Leaf yellowing and/or withering	*X. c.* pv. *oryzae*
Lolium	Leaf streaking and wilting	*X. c.* pv. *graminis*
Lycopersicon	Wilting of leaves	*Clavibacter michiganense* subsp. *michiganense* *Pseudomonas corrugata* *P. solanacearum*
	Wilting of whole plant	*P. solanacearum* *P. corrugata*
	Irregular yellow-brown necrotic areas on leaves	*Clavibacter michiganense* subsp. *michiganense* *Xanthomonas campestris* pv. *vesicatoria*
	Mealiness of leaves/stems	*Clavibacter michiganense* subsp. *michiganense*
	Scorching of leaves	*C. michiganense* subsp. *michiganense*
	Cankering	*C. michiganense* subsp. *michiganense* *Xanthomonas campestris* pv. *vesicatoria*
	Yellow discoloration of vascular system	*Clavibacter michiganense* subsp. *michiganense*
	Brown discoloration of vascular system	*Pseudomonas solanacearum* *P. corrugata*
	Pith necrosis	*P. corrugata*
	Marbling of young green fruit	*Clavibacter michiganense* subsp. *michiganense*
	Spotting of fruit	*Xanthomonas campestris* pv. *vesicatoria* *Clavibacter michiganense* subsp. *michiganense* *Pseudomonas syringae* pv. *tomato*
	Production of adventitious roots	*P. solanacearum* *P. corrugata*
	Leaf and stem spotting and necrosis	*Xanthomonas campestris* pv. *vesicatoria* *Pseudomonas syringae* pv. *tomato* *P. cichorii*

Host genus	Symptom	Possible pathogen(s)
		P. marginalis *P. viridiflava*
	Stem necrosis	*P. marginalis* *P. viridiflava*
Mallotus	Leaf spotting	*Erwinia mallotivora*
	Shoot blight	*E. mallotivora*
Malus	Root proliferation on stems or roots	*Agrobacterium rhizogenes*
	Rounded galls on stem, usually at soil level or on root	*A. tumefaciens*
	Blossom blighting	*Erwinia amylovora* *Pseudomonas syringae* pv. *syringae*
	Branch cankering	*Erwinia amylovora* *Pseudomonas syringae* pv. *syringae*
	Dieback of shoots	*Erwinia amylovora* *Pseudomonas syringae* pv. *syringae*
	Shoot wilting	*Erwinia amylovora*
	Fruit rotting	*E. amylovora*
	Fruit blistering	*Pseudomonas syringae* pv. *papulans*
Mangifera	Leaf spotting	*Xanthomonas campestris* pv. *mangiferaeindicae*
	Fruit spotting	*X. c.* pv. *mangiferaeindicae*
	Stem cankering	*X. c.* pv. *mangiferaeindicae*
Manihot	Leaf spotting	*X. c.* pv. *manihotis* *X. c.* pv. *cassavae*
	Wilting and shoot dieback	*X. c.* pv. *manihotis*
Matthiola	Leaf necrosis	*X. c.* pv. *campestris*
	Veinal necrosis	*X. c.* pv. *campestris* *X. c.* pv. *incanae*
	Leaf yellowing	*X. c.* pv. *campestris* *X. c.* pv. *incanae*
Medicago	Leaf and stem spotting	*X. c.* pv. *alfalfae* *Pseudomonas marginalis*
	Damping off	*Xanthomonas campestris* pv. *alfalfae*
	Wilting and stunting	*Clavibacter michiganense* subsp. *insidiosum*

Host genus	Symptom	Possible pathogen(s)
Morus	Rounded galls on stem, usually at soil level or on root	*Agrobacterium tumefaciens*
	Leaf spotting	*Pseudomonas syringae* pv. *mori*
	Cankering	*P. s.* pv. *mori*
Mucuna	Leaf spotting	*P. andropogonis*
Musa	Wilting, necrosis	*P. solanacearum*
Nerium	Stem cankering and leaf galling	*P. syringae* subsp. *savastanoi*
Nicotiana	Leaf spotting	*P. s.* pv. *tabaci* *P. cichorii* *P. marginalis*
	Wilting	*P. solanacearum*
Olea	Stem cankering, and knotting	*P. s.* subsp. *savastanoi*
Oryza	Leaf and stem spotting	*P. avenae* *P. s.* pv. *oryzicola*
	Leaf striping	*P. avenae* *Xanthomonas campestris* pv. *oryzicola* *X. c.* pv. *oryzae*
	Leaf yellowing and/or withering	*X. c.* pv. *oryzae* *X. c.* pv. *oryzicola*
Panicum	Stunting, yellowing, wilting	*Erwinia chrysanthemi*
	Leaf striping	*Pseudomonas avenae*
Papaver	Leaf necrosis	*P. cichorii*
Paspalum	Leaf striping	*P. rubrilineans*
	Top rotting	*P. rubrilineans*
Pastinaca	Necrosis, rotting	*P. marginalis*
Pelargonium	Leaf spotting	*Xanthomonas campestris* pv. *pelargonii*
	Leaf zonal necrosis	*X. c.* pv. *pelargonii*
	Stem rotting	*X. c.* pv. *pelargonii*
	Leafy galling	*Rhodococcus fascians*
Pennisetum	Stunting, yellowing, wilting	*Erwinia chrysanthemi*
	Leaf blighting	*Pseudomonas syringae* pv. *syringae*
Persea	Leaf and shoot necrosis	*P. s.* pv. *syringae*
Phaseolus	Leaf, stem and pod spotting	*P. s.* pv. *phaseolicola* *P. s.* pv. *syringae*

Host genus	Symptom	Possible pathogen(s)
		P. viridiflava
		Xanthomonas campestris pv. *phaseoli*
		Curtobacterium flaccumfaciens pv. *flaccumfaciens*
	Leaf mosaic and malformation	*P. s.* pv. *phaseolicola*
	Leaf necrosis	*P. s.* pv. *phaseolicola*
		P. s. pv. *syringae*
		P. viridiflava
		Xanthomonas campestris pv. *phaseoli*
	Stem galling	*Pseudomonas viridiflava*
	Wilting	*Curtobacterium flaccumfaciens* pv. *flaccumfaciens*
Philadelphus	Leaf spotting	*P. syringae* pv. *philadelphii*
Philodendron	Rotting of leaves and stems	*Erwinia chrysanthemi*
Phleum	Leaf streaking	*Xanthomonas campestris* pv. *phlei*
Physalis	Leaf spotting	*X. c.* pv. *vesicatoria*
Piper	Leaf and shoot necrosis	*Pseudomonas syringae* pv. *syringae*
Pisum	Leaf stem and pod spotting	*P. s.* pv. *pisi*
		P. s. pv. *syringae*
		P. viridiflava
Poa	Leaf spotting	*Xanthomonas campestris* pv. *poae*
Populus	Swelling or cankering on shoots and branches	*X. populi* pv. *populi*
	Bark necrosis	*X. campestris* pv. *populi*
	Dieback	*Erwinia salicis*
Primula	Leaf spotting	*Pseudomonas syringae* pv. *syringae*
		P. s. pv. *primulae*
		P. viridiflava
Prunus	Rounded galls on stem, usually at soil level or on root	*Agrobacterium tumefaciens*
	Cankering on branches and shoots	*Pseudomonas amygdali*
		P. syringae pv. *syringae*

Host genus	Symptom	Possible pathogen(s)
		P. syringae pv. *morsprunorum* *Xanthomonas campestris* pv. *pruni*
	Swellings on twigs and branches	*Pseudomonas amygdali* *Xanthomonas campestris* pv. *pruni*
	Leaf spotting	*Pseudomonas syringae* pv. *morsprunorum* *P. s.* pv. *syringae* *P. s.* pv. *persicae* *Xanthomonas campestris* pv. *pruni*
	Fruit spotting	*X. c.* pv. *pruni*
	Dieback of shoots	*Pseudomonas syringae* pv. *syringae* *P. s.* pv. *morsprunorum* *P. s.* pv. *persicae* *Xanthomonas campestris* pv. *pruni*
	Bud death	*Pseudomonas syringae* pv. *syringae* *P. s.* pv. *morsprunorum*
	Blossom blight	*P. s.* pv. *syringae*
Pyracantha	Blossom blight, leaf and shoot necrosis	*Erwinia amylovora*
Pyrus	Root proliferation on root or stem	*Agrobacterium rhizogenes*
	Rounded galls on stem usually at soil level or on root	*A. tumefaciens*
	Blossom blight	*Erwinia amylovora* *Pseudomonas syringae* pv. *syringae*
	Branch cankering	*Erwinia amylovora* *P. s.* pv. *syringae*
	Dieback of shoots	*E. amylovora* *P. s.* pv. *syringae*
	Shoot wilting	*E. amylovora*
	Fruit rotting	*E. amylovora* *Gluconobacter oxydans*
Quercus	Exudation from acorns and cups	*Erwinia quercina*
Raphanus	Leaf necrosis	*Xanthomonas campestris* pv. *campestris*

Host genus	Symptom	Possible pathogen(s)
	Veinal necrosis	*X. c.* pv. *campestris*
	Leaf yellowing	*X. c.* pv. *campestris*
Rheum	Crown rotting	*Erwinia rhapontici*
Ricinus	Leaf spotting and necrosis	*Xanthomonas campestris* pv. *ricini*
	Stem necrosis	*X. c.* pv. *ricini*
Rosa	Root proliferation on root or stem	*Agrobacterium rhizogenes*
	Rounded galls on stem mainly at soil level or on roots	*A. tumefaciens*
	Leaf and shoot necrosis	*Pseudomonas syringae* pv. *syringae*
Roystonea	Leaf spotting and/or striping	*Xanthomonas campestris* pv. *vasculorum*
Rubus	Root proliferation on root or stem	*Agrobacterium rhizogenes*
	Rounded galls on stem mainly at soil level or on roots	*A. tumefaciens* *A. rubi*
	Erumpent longitudinal galls on canes	*A. tumefaciens*
Saccharum	Leaf mottling	*Erwinia chrysanthemi*
	Leaf mosaic	*E. chrysanthemi*
	Leaf striping	*Pseudomonas rubrilineans* *P. rubrisubalbicans*
	Top rotting	*P. rubrilineans* *Xanthomonas campestris* pv. *vasculorum*
	Leaf scalding	*X. albilineans* *X. campestris* pv. *vasculorum*
	Yellow slime	*X. c.* pv. *vasculorum*
Saintpaulia	Rotting of leaves and petioles	*Erwinia chrysanthemi* *E. carotovora*
Salix	Rounded galls on stem, usually at soil level, or on root	*Agrobacterium tumefaciens*
	Wilting	*Erwinia salicis*
	Reddening of leaves	*E. salicis*
Secale	Leaf and stem spotting	*Xanthomonas campestris* pv. *secalis*

Host genus	Symptom	Possible pathogen(s)
Sesamum	Leaf spotting	*Pseudomonas syringae* pv. *sesami* *Xanthomonas campestris* pv. *sesami*
	Stem lesions	*Pseudomonas syringae* pv. *sesami* *Xanthomonas campestris* pv. *sesami*
	Capsule blackening	*Pseudomonas syringae* pv. *sesami* *Xanthomonas campestris* pv. *sesami*
Setaria	Leaf striping	*Pseudomonas avenae*
Solanum	Stunting	*Erwinia carotovora* *E. chrysanthemi*
	Wilting	*E. carotovora* *E. chrysanthemi* *Pseudomonas solanacearum* *Clavibacter michiganense* subsp. *sepedonicum*
	Chlorosis of leaves, leaf rolling	*E. carotovora* *E. chrysanthemi* *Clavibacter michiganense* subsp. *sepedonicum*
	Stem lesions	*Erwinia carotovora* *E. chrysanthemi* *Pseudomonas solanacearum*
	Tuber rotting	*Erwinia carotovora* *Clavibacter michiganense* subsp. *sepedonicum* *Pseudomonas solanacearum* *P. marginalis* *P. viridiflava*
Sorbus	Blossom blight, leaf and shoot necrosis	*Erwinia amylovora*
Sorghum	Leaf spotting	*Pseudomonas andropogonis* *P. syringae* pv. *syringae*
	Leaf striping	*P. andropogonis* *P. syringae* pv. *syringae* *Xanthomonas campestris* pv. *holcicola*

Host genus	Symptom	Possible pathogen(s)
Stranvaesia	Blossom blight, leaf and shoot necrosis	*Erwinia amylovora*
Syringa	Leaf necrosis	*Pseudomonas syringae* pv. *syringae*
	Dieback of shoots	*P. s.* pv. *syringae*
Tagetes	Leaf spotting	*P. syringae* pv. *tagetis*
Thysanolaena	Leaf spotting and/or striping	*Xanthomonas campestris* pv. *vasculorum*
Trifolium	Leaf and stem spotting	*Pseudomonas andropogonis* *P. syringae* pv. *syringae* *Xanthomonas campestris* pv. *alfalfae*
Triticum	Yellow slime on upper parts	*Clavibacter tritici*
	Yellow or white spots on leaves and sheaths	*C. tritici*
	Pink grain	*Erwinia rhapontici*
	Leaf spotting and streaking	*Pseudomonas syringae* pv. *atrofaciens* *P. s.* pv. *syringae* *Xanthomonas campestris* pv. *undulosa*
	Glume blotching	*Pseudomonas syringae* pv. *atrofaciens* *Xanthomonas campestris* pv. *undulosa*
	Seed blackening	*Pseudomonas syringae* pv. *atrofaciens*
Tulipa	Spotting and cracking of leaves	*Curtobacterium flaccumfaciens* pv. *oortii*
	Stunting and veinal silvering	*C. flaccumfaciens* pv. *oortii*
	Spotting and necrosis of scales	*C. flaccumfaciens* pv. *oortii*
Vicia	Leaf spotting	*Pseudomonas andropogonis* *P. syringae* pv. *pisi* *P. s.* pv. *syringae*
	Stem lesions	*P. andropogonis* *P. s.* pv. *pisi* *P. s.* pv. *syringae*
Vigna	Leaf spotting	*Pseudomonas syringae* pv. *pisi* *P. s.* pv. *syringae* *P. viridiflava*

Host genus	Symptom	Possible pathogen(s)
	Stem and pod lesions	*P. s.* pv. *pisi* *P. s.* pv. *syringae*
Vitis	Leaf spotting	*P. s.* pv. *syringae* *P. viridiflava* *Xanthomonas ampelina*
	Dieback	*Pseudomonas syringae* pv. *syringae* *Xanthomonas ampelina*
	Blighting	*X. ampelina*
	Stem galling	*Agrobacterium tumefaciens*
Zea	Stunting and wilting	*Erwinia chrysanthemi* *E. stewartii* *Clavibacter michiganense* subsp. *nebraskense*
	Necrosis or soft rotting of leaves	*Erwinia chrysanthemi*
	Leaf striping	*E. stewartii* *Xanthomonas campestris* pv. *vasculorum* *Pseudomonas avenae* *P. andropogonis* *P. rubrilineans* *Clavibacter michiganense* subsp. *nebraskense* *Xanthomonas campestris* pv. *holcicola*
	Top death	*Erwinia stewartii* *Pseudomonas avenae* *P. rubrilineans* *Xanthomonas campestris* pv. *vasculorum*
	Stalk rotting	*Erwinia chrysanthemi* *Pseudomonas avenae*
	Root rotting	*Clavibacter michiganense* pv. *nebraskense*
	Seedling blight	*Erwinia stewartii*
Zingiber	Wilting	*Pseudomonas solanacearum*
Zinnia	Leaf spotting	*Xanthomonas campestris* pv. *zinniae*
Zizania	Leaf striping	*X. c.* pv. *oryzae*
	Leaf yellowing and/or withering	*X. c.* pv. *oryzae*

3

Diagnostic Procedures for Bacterial Plant Diseases

3.1 Introduction

This chapter gives methods for the isolation of plant pathogenic bacteria and presumptive and confirmatory diagnoses of the diseases they cause. The schemes presented take into account the need for rapid diagnosis often demanded of the diagnostician. The diagnostic and identification schemes used are based largely upon nutritional and physiological tests which, in experienced hands, are reliable and give reproducible results.

In the main, non-selective rather than selective isolation media are recommended because they usually simplify diagnosis and lead to more rapid and accurate results, provided suitable diseased tissue is selected for isolation and bacteria are demonstrated microscopically in large numbers within that tissue (see 1.3). However, selective media are useful in special circumstances: for example, the isolation of plant pathogenic bacteria from soil.

The presumptive diagnoses given in this chapter rely on the evidence of symptom expression and the minimum number of tests required to give rapid but reasonably accurate answers. Confirmation usually requires further tests including a host test, and although the inexperienced worker may need to confirm diagnoses frequently, the more experienced worker should be able to rely upon presumptive diagnoses for many cases, especially when advice or action is urgently required. It is important, however, to stress that even experienced diagnosticians should confirm a proportion of their presumptive diagnoses, particularly if symptoms are not typical. This is because such diagnoses involve, to a greater or lesser degree, subjective judgement, and particularly because one's conception of the symptom picture for particular diseases can imperceptibly but erroneously shift with time unless checked. Such a shift can have a variety of causes, but commonly it occurs when the diagnostician designates symptoms similar to, but slightly different from, those he/she normally associates with a particular disease as being atypical of that disease. If he/she sees enough of these 'atypical' symptoms they tend to become incorporated into the symptom picture for that disease even though they may be those of a similar but distinct disease.

The literature reports more rapid diagnostic methods than those given here for many diseases, but they have not usually been tried and tested by general use, are not always reliable, the results may be difficult to interpret in inexperienced hands and they may require equipment not readily available in the diagnostic laboratory.

3.2 Hyperplasias

Hyperplasia in the context of plant disease refers to an overgrowth of tissue produced by rapid cell multiplication. This may result in the formation of a gall on almost any part of the plant or may result in a more organized over-production or proliferation of certain organs, e.g. roots, stems or flowers, often as a classical fasciation with the organs in question fused together side by side. Galls may consist primarily of parenchyma tissue showing little differentiation, as in crown gall, or may consist of more organized and recognizable tissues as in bud proliferation (leafy gall) caused by *Rhodococcus fascians* and root proliferation (hairy root) caused by *Agrobacterium rhizogenes*.

The inciting bacteria in galls, especially those caused by *Agrobacterium* spp. and *Rhodococcus fascians*, are usually few in number and are generally confined to the outer cell layers of the gall.

3.2.1 Crown gall (*Agrobacterium tumefaciens* and *A. rubi*)

The nomenclature of the genus *Agrobacterium* is confused. There are three biochemically distinct groups (biovars) within the genus that equate taxonomically with three separate species. However, it is not possible to name these biovars as species because of the existing nomenclature. Thus, *A. tumefaciens* comprises tumorigenic strains belonging to biovars I, II and III; *A. radiobacter*, a soil inhabiting saprohyte, comprises non-tumorigenic strains of biovars I, II and III; *A. rhizogenes* comprises rhizogenic strains of biovars I and II; and *A. rubi* comprises a few strains unrelated at species level to the three biovars. Pathogenicity is determined by plasmids that can be readily transferred to other, virulent and avirulent, agrobacteria.

This is an extremely common, often serious, disease with a very wide host range. Strains usually show some host specificity, but the full host range of different strains and the extent to which ranges overlap is not known. It is economically important on members of the Rosaceae including stone fruits, pome fruits, *Rubus* spp. and also on chrysanthemums and grapevine.

Symptoms. On most hosts leaves are unaffected, although occasionally galls are formed on petioles. Galls can be common on leaf surfaces of chrysanthemums grown under glass.

Stems have initially neat, rounded, smooth or fissured galls, usually just above soil level. On perennial woody hosts galls became more woody and fissured with age, can be more than 10 cm across and can girdle the stem. In a few hosts, erumpent galls can burst through the outer stem tissues forming a continuous mass of gall tissue extending along the stem for up to 1 m or more; this symptom is common on blackberry and raspberry and known as 'cane gall'.

Roots have rounded galls similar to those on stems but usually much smaller. For further information on symptoms and host ranges see Wormald (1945), De Cleene and De Ley (1976).

Isolation. Wash the gall in tapwater, then slice pieces of young, fresh tissue from the gall surface. Crush in a few drops of sterile water in a sterile Petri dish. If the gall is old, rough or contaminated with soil, clean it thoroughly by scrubbing gently, immerse momentarily in a suitable wetting agent (e.g. 0.1% Manoxol), surface sterilize by soaking in hypochlorile solution (0.5% available chlorine) for 10 minutes and rinse thoroughly in three changes of sterile water. The crushed gall should be left for at least 20 minutes and preferably for several hours to allow the bacteria to diffuse out of the tissues; diffusion for 6 hours is recommended with older galls.

Streak the suspension on plates of nutrient dextose agar (6.2.10) or potato dextrose agar (6.2.13) and on the semi-selective Clark's medium (6.2.3). If it is desired to determine the biovars that may be present, also streak on the following selective media: either Schroth's medium (6.2.14), Kerr's medium (6.2.6) and Roy and Sasser's medium (6.2.23) *or* on Brisbane and Kerr's media 1A, 2E and 3DG (6.2.2).

Examine plates of the non-selective medium after 2–3 days and selective media after 4–5 days. On all media, colonies of agrobacteria are neat, round, entire, smooth, domed and mucoid. On the selective media biovars will be isolated as in Table 3.1.

Comparison with plates streaked with cultures of known biovars is advisable until colony morphology on the medium or media of choice is recognized easily. A presumptive diagnosis can be made at this stage (see below).

Test procedures. If investigating colonies from selective media purify them by re-streaking separate water suspensions of each of 2–3 colonies thought to be of *A. tumefaciens* on plates of non-selective media. Prepare water suspensions from each of 2–3 such 'purified' colonies and from each inoculate lactose yeast extract plates (6.3.7), O/F medium (6.3.1) and a range of young host plants by stabbing into stems or inoculating root slices (5.3.7.2). Sunflower, marigold, tomato and bryophyllum

Table 3.1. Isolation of biovars on selective media

Medium	Selective for biovar	Semi-selective for biovar	Colour
Schroth	I		Yellow/tan
Kerr	II		Pearly white/tan
Clark		I, II, III	Pearly white
Brisbane & Kerr medium 1A	I		
Brisbane & Kerr medium 2E	II		White, often becoming brown in older cultures
Brisbane & Kerr medium 3DG	III		
Roy & Sasser medium 3	III		Pinkish-white (red centre)

Biovar I grows faster than biovars II and III.

(*Kalanchoe* sp.) are useful test plants because they react to most strains from herbaceous plants. If plants of the same host as that diseased are available, inoculate these.

If investigating colonies from non-selective media, omit the purification step but, in addition to inoculating the media and test plants given above, re-streak each suspension on a non-selective plate. If the growth on a plate indicates that the suspension from which it was streaked derived from a colony on the isolation plate that was a mixture of different bacteria ignore the results on media inoculated with that suspension.

After 2 days, test the lactose yeast extract plates for 3-keto lactose and after 2–3 days read the O/F test. Most biovar I agrobacteria produce 3-keto lactose; biovar II and III isolates do not. All agrobacteria are oxidative. Most strains of agrobacteria from crown galls will produce galls on tomato, sunflower, marigold or bryophyllum within 10–14 days, but it is often difficult to demonstrate pathogenicity of isolates from grape and from apple and other woody rosaceous plants. Observe plants for gall formation for up to 28 days. Biovar III isolates have been isolated almost exclusively from grape vine and chrysanthemum.

Presumptive diagnosis. An experienced diagnostician can make a presumptive diagnosis if symptoms are typical and if Gram-negative bacteria with typical colony characters are dominant on non-selective or selective media. For those less experienced or where there is doubt, at least confirm that typical colonies are oxidative; preferably attempt to confirm the diagnosis with a host test.

Confirmation of diagnosis. Confirm the diagnosis if typical symptoms are attained in a host test (see above and 5.3.7.2). For further information on the taxonomy of agrobacteria see Kersters and De Ley (1984); on the disease, see Panagopoulos (1987).

3.2.2 Hairy root disease, cucumber root mat (*Agrobacterium rhizogenes*)

Symptoms. In hairy root disease a very large number of small roots, fleshy at first and fibrous later, are produced that protrude from stems or roots or from a localized hard swelling or graft unions. In roses they can also arise from the ends of cuttings or at disbud scars. In cucumber mat disease a mass of fine roots forms at or on the soil surface.

Isolation. The fine roots should be carefully washed in water to remove all soil debris. Do not surface sterilize. Crush the cleaned roots in a few drops of sterile water. Leave to stand in a Petri dish or bijou bottle for approximately 6 hours. Streak on isolation plates as for crown gall (3.2.1).

Test procedures. As for crown gall (3.2.1).

Presumptive diagnosis. As for crown gall (3.2.1)

Confirmation of diagnosis. Confirm the diagnosis if rootlets are produced on slices of freshly harvested carrot or beetroot (5.3.7.2). Most if not all *A. rhizogenes* isolates are biovar II except some of those that cause cucumber root mat and which are biovar I. For further information on symptoms and host range see Munnecke *et al.* (1963) and De Cleene and De Ley (1981); for information on the taxonomy of agrobacteria see Kersters and De Ley (1984).

3.2.3 Leafy gall (*Rhodococcus fascians* syn. *Corynebacterium fascians*)

A common and widespread disease in Europe and North America, of some economic importance in the production of ornamentals from cuttings.

Symptoms. Many buds are produced, usually at ground level, that develop into a mass of short, thick, distorted stems with misshapen leaves. Outgrowths may sometimes resemble galls. The 'cauliflower' effect of the crown of the plant in strawberry is also associated with nematode infection. Sometimes fasciation results in a witch's broom effect. These symptoms are due to loss by the host of apical dominance due to interference by the bacterium with the plant's hormonal control. Blossom can be distorted with multiple petals; this symptom is not common.

Isolation. Excise clean, fresh buds or young tip tissue and crush in a few drops of sterile water. Allow to stand for at least 20 minutes and preferably 2–3 hours. Streak the suspension on nutrient dextrose agar (6.2.10) or yeast glucose mineral agar (6.2.21). A dichromate selective medium is sometimes recommended but it seems of no advantage except where heavy contamination with other bacteria is expected.

R. fascians grows slowly and after 3–5 days colonies are small, deep yellow to bright orange, convex, smooth, shiny, entire and opaque.

Test procedures. Prepare Gram stains from likely colonies. *R. fascians* is a Gram-positive pleomorphic rod, usually longer than the other plant pathogenic coryneforms. Coryneform cells, palisade and Chinese letter arrangements are often seen but the majority of cells are found singly or in pairs. Old colonies may stain weakly or occasionally appear Gram-negative or Gram-variable. Check microscopically that colonies are not apparent mixtures of bacteria. Two or three typical unmixed colonies should each be transferred to separate slopes of Christensen's urea medium (6.4.2).

Presumptive diagnosis. A presumptive diagnosis of leafy gall can be made if typical Gram-positive colonies which are slow-growing but give a positive urease reaction are isolated from plants with typical symptoms.

Confirmation of diagnosis. Obtain purified cultures by restreaking on nutrient dextrose medium, pick well-separated colonies and from these inoculate sweet pea seedlings (5.3.7.3). Sweet pea seedlings take about 7 days to raise for inoculation and should be set up when original isolations are made. A few strains do not infect sweet pea. If such isolates are suspected inoculate plants of the host from which cultures were isolated or failing that, gram (*Cicer arietinum*) seedlings (5.3.7.3).

Confirm the diagnosis of leafy gall if typical symptoms are obtained in a host test.

3.2.4 Ash canker, olive knot, oleander knot (*Pseudomonas syringae* subsp. *savastanoi* pathovars *fraxinii*, *olei* and *nerii* respectively)

The biology and taxonomy of this group of pathogens has recently been discussed and reviewed by Janse (1982).

The disease is common and often limiting in many olive-producing areas, and common and often serious on ash in north-west Europe.

Symptoms. On olive and oleander, leaves are sometimes affected. Petioles and midribs may show small galls or tubercles causing premature abscission. Stems develop hyperplastic galls or knots develop on young twigs which may be killed as a result; galls are at first small then increase in size to become irregular, sponge-like, cracked and fissured. On ash the most common symptom is an erumpent hyperplastic stem canker. Trees are often grossly disfigured.

Isolation. Excise young, undecayed areas of the galls or knots. If the tissue is woody, scrape the surface, flame with alcohol and crush with a scalpel in a few drops of sterile water. If the tissue is soft, wipe with alcohol and crush in a few drops of sterile water. Allow to stand for 2–3 hours. Streak the suspension on to 5% sucrose nutrient agar (6.2.17) and King's medium B (6.2.7).

Test procedures. If after 48 hours, levan (white, smooth, high-domed, mucoid) colonies are absent from the 5% sucrose nutrient agar plates and fluorescent colonies are numerous on the King's medium B plate, streak another King's medium B plate and inoculate LOPAT test media (see 3.3.2 and Table 3.2) from a suspension in water of a single representative colony from the King's B medium. Repeat for at least one other colony. If more than one fluorescent colony type is seen inoculate a set of LOPAT test media for each. Incubate for 2–3 days and read the results of the LOPAT tests (see 3.3.2 and Table 3.2) at the appropriate times. If the streaked King's medium B plate is clearly a mixture ignore the results of LOPAT tests from that colony. If levan colonies are present on the 5% sucrose nutrient agar isolation plates, streak fluorescent colonies from the King's medium B plates on 5% sucrose nutrient agar plates and then carry out LOPAT tests for any 'non-levan' cultures.

Presumptive diagnosis. A presumptive diagnosis for these diseases can be made if green-fluorescent isolates with the LOPAT characters − − − − ± are obtained from affected tissue of plants with typical symptoms.

It should be noted that not all isolates are non-levan; some are intermediate in colony type between the levan and non-levan type and a few indistinguishable from the levan type. Variability of olive isolates is discussed by Ercolani (1983). The tobacco reaction can also be variable. The main determinative problem is to differentiate the atypical levan isolates of *P. syringae* subsp. *savastanoi* from *P. syringae* subsp. *syringae* which may present as contaminants on the galls. *P. syringae* subsp. *savastanoi*, unlike most *P. syringae* subsp. *syringae* isolates, does not liquefy gelatine or utilize quinate, erythritol and L-lactate as sole carbon sources (6.3.2).

Confirmation of diagnosis. Confirm the diagnosis of the relevant disease if typical symptoms are obtained in a host test (5.3.7.1). For further information on the disease in ash see Janse (1981a and b) and Boa and Preece (1981).

3.3 Leaf, stem and fruit spots and necroses

The pathogens causing leaf, stem and fruit spots and necroses fall into four main taxonomic groups: the green-fluorescent pseudomonads; the non-fluorescent pseudomonads; the xanthomonads; and the coryneforms. Many causing this type of symptom are pathovars of *Pseudomonas syringae* or *Xanthomonas campestris*. Within each of these two species there are as

Table 3.2. Grouping of green-fluorescent pseudomonads by the LOPAT scheme (Lelliott *et al.*, 1966)

	Presumptive (LOPAT) characters					Confirmatory characters				
Group	Levan type colonies (6.2.17)	Oxidase reaction (6.6.2)	Potato rot (5.3.6) or pectate gel liquefaction	Arginine dihydrolase (6.4.3)	Tobacco hyper-sensitivity (5.2.1)	2-keto gluconate production (6.3.6)	Egg yolk reaction* (6.5.1)	Nitrate reduction (6.4.1)	Acid from sucrose (6.3.3)	Species
Ia	+	–	–	–	+	–	–	–	+	*P. syringae*
Ib	–	–	–	–	+	–	–	–	+	*P.s.* subsp. *savanastoi*; *P. delphini*
II	–	–	+	–	+	–	–	–	–	*P. viridiflava*
III	–	+	–	–	+	–	–	d	–	*P. cichorii*; *P. agarici*
IVa	+	+	+	+	–	+	+	+	+	*P. marginalis* (pectolytic *P. fluorescens*)
IVb	–	+	+	+	–	+	d	d	d	*P. fluorescens*
Va	–	+	–	+	–	+	+	–	–	*P. tolaasii*, and some saprophytic pseudomonads
Vb	+	+	–	+	–	+	d	d	d	*P. fluorescens* and some other saprophytic pseudomonads

d = result varies with isolate.

*Although Lelliott *et al.* (1966) originally used the lipolysis of margarine as one of the confirmatory characters, difficulties can arise due to delayed positives or delayed recognition of positives. As the egg yolk reaction correlated very closely with the margarine lipolysis test and there was generally no difficulty in interpretation of the results, the egg yolk can replace the lipase test as a confirmatory character.

yet no reliable and well-proven means, other than host tests, for differentiating them. Even using host tests, differentiation at pathovar level is not always certain since, although some pathovars are host specific to a single host, others have several or many hosts and there is some overlapping of the host ranges of different pathovars. Also, the host range of some is not known with certainty and the nature of many host tests are such that they can yield erroneous positive results unless used with care and with adequate positive and negative controls.

Most of the green-fluorescent pathogenic pseudomonads produce diffusible pigments on King's medium B (6.2.7) that fluoresce green to blue in ultraviolet light and are usually formed more slowly and in smaller amounts than the similar pigments of most related saprophytic bacteria. Occasionally atypical strains of group 1 (see below) that do not produce fluorescent pigments are isolated, but strains of *P. syringae* pv. *morsprunorum* often do not and *P. syringae* pv. *persicae* and race 1 of *P. syringae* pv. *pisi* do not produce them. Several pathogens also produce other disease symptoms, e.g. *P. syringae* pv. *syringae*, *P. syringae* pv. *morsprunorum* and *Xanthomonas campestris* pv. *citri* produce cankers as well as leaf spots.

The use of the LOPAT scheme (Table 3.2) will enable separation of pathogenic from saphrophytic fluorescent pseudomonads from plants and will group the pathogens. It should be noted that no saprophytes are known with the characters of groups I, II, III and IVa. Group V are all saprophytes of the *P. fluorescens-putida* complex with the exception of *P. tolaasii* which has the characters of group Va. Results apply only if read after 3 days incubation with the exception of oxidase, which should be read at 1–2 days, potato rot at 2 days and tobacco reaction at 2–5 days. Pectolytic *Pseudomonas* spp. cause only a shallow crater round discrete colonies compared with pectolytic *Erwinia* spp. (other than those of the non-soft rotting section) that cause deep craters and/or liquefaction.

There are a number of determinative schemes that claim to differentiate some at least of the *Pseudomonas syringae* pathovars. These should be treated cautiously, since it is probably fair to say that the differences in interpretation and reproducibility of such schemes between different diagnostic laboratories are greater than between the pathovars themselves. This is probably equally true for many nutritional, physiological, serological and 'phage specificity tests. So far as we are aware, none have been proven by widespread use in diagnostic laboratories. The schemes presented here will differentiate between pathogens and non-pathogenic bacteria found in plant tissue and will give a reasonably accurate diagnosis to species, together with a fairly accurate indication of the pathovar in most cases.

3.3.1 Preliminary schedule for the diagnosis of leaf, stem and fruit spots and necroses

Symptoms. If the symptoms are clearly those of a particular disease, or of only a very few diseases, refer to the relevant sections in which these are dealt with. If the symptoms are not attributable in this way proceed as follows:

Isolation. Examine suspensions of diseased tissue microscopically (1.3) and streak suspensions containing bacteria (1.4) on plates of 5% sucrose nutrient agar (6.2.17), King's medium B (6.2.7) and nutrient dextrose agar (6.2.10). Incubate for 2–3 days and if the dominant organism on the plates corresponds to one of the descriptions in Table 3.3, proceed to the test procedures of the relevant section of this chapter; if organisms with the colony forms of more than one pathogen are present, investigate each type independently. However, it is advisable for the diagnostician, particularly if inexperienced, to determine the Gram reaction of bacteria suspected of being the pathogen and whether they are oxidative or fermentative as he/she proceeds. This can enable the preliminary judgement based on Table 3.3 to be corrected quickly since all pseudomonads and xanthomonads are Gram-negative and oxidative, all coryneforms are Gram-positive and oxidative, all erwinias are Gram-negative and fermentative and most bacilli are Gram-positive and all are fermentative.

3.3.2 Green-fluorescent pseudomonads

If consideration of the results of isolation plates (3.2.1 and Table 3.3) indicates that the pathogen likely to be involved is a green-fluorescent pseudomonad, inoculate sets of LOPAT test media (see below) and re-streak on plates of King's medium B from separate, water suspensions of each of 2–3 well-isolated, levan-type colonies and/or green-centred, non-levan, smooth colonies if they are present on the 5% sucrose nutrient agar isolation plate. If neither colony type is present inoculate a set of LOPAT test media and re-streak King's medium B from each of 2–3 suspensions of well-isolated green-fluorescent colonies on the King's B isolation plate. If the 5% sucrose nutrient agar isolation plates are too overgrown to distinguish isolated colonies, inoculate sets of LOPAT test media from green-fluorescent colonies in a similar way to that above but additionally streak each of the 2–3 colonies on 5% sucrose nutrient agar plates.

LOPAT test media are:

1 5% sucrose nutrient plate (6.2.17).

Table 3.3. Characteristics associated with colonies of pathogens on isolation plates

5% sucrose nutrient agar	King's B	Nutrient dextrose agar	Possible pathogen
Whitish, domed, smooth, mucoid, (levan)	Whitish-grey, raised with diffusible yellowish green pigment that fluoresces blue-green under ultra violet light	Whitish-grey, raised, butyrous	Green-fluorescent pseudomonad of group Ia, IVa
Whitish-grey, raised	As above	As above	Green-fluorescent pseudomonad of group Ib, III, IVb
Convex, smooth, whitish to yellowish with greenish centre	As above	As 5% SNA	Green-fluorescent pseudomonad of group II
Convex, smooth or wrinkled, whitish-green, to yellowish brown; diffusible or non-diffusible	Whitish-grey but often producing pigments which do not fluoresce under ultraviolet light	Convex, smooth or wrinkled, whitish-green, yellowish to brown; diffusible or non-diffusible	Non-fluorescent pseudomonad

pigments may be produced		pigments may be produced	
High convex to domed, smooth, creamy to yellowish	As 5% SNA. No diffusible fluorescent pigment	High convex to domed smooth, mucoid, creamy to yellow; brown diffusible pigment produced rarely	Xanthomonads
Raised, convex or domed, smooth mucoid (occasional levan production[†]) usually white to whitish-grey	Raised or convex, whitish-grey which do not fluoresce under ultraviolet light	Raised or convex, whitish grey	Erwinia
Convex-domed, smooth, mucoid, yellowish-orange[‡]	Raised or convex, smooth, white-yellowish[‡]	Raised, convex to domed, mucoid, yellowish-orange[‡]	Coryneform

*Occasionally atypical, non-fluorescent forms of some pathogens of Group I are isolated; but they are common in *P. syringae* pv. *morsprunorum*, and are the rule in *P. s.* pv. *persicae*.
†Erwinia amylovora produces whitish, domed, smooth, mucoid levan colonies.
‡Gram positive. Pink, blue or violet pigments sometimes produced.

2 Nutrient agar slope (6.2.9).
3 Potato slice (5.3.6) or pectate gel plate (6.2.12).
4 Arginine medium (6.4.3).

The plates should be streaked to obtain separate colonies. After 2 days' incubation examine the re-streaked King's medium B plates. If any are obviously mixtures of bacteria ignore any results from the LOPAT tests made from that colony. For any apparently unmixed culture proceed as follows. After 2 days test the growth from the nutrient agar slopes for oxidase reaction and use water suspensions of growth from them to inoculate tobacco plants for hypersensitivity (5.2.1). After 2 days, test potato slices for rot or record pectinase activity and record fluorescence on King's medium B slopes. At 3 days record arginine reaction.

From the LOPAT results determine the group(s) isolated (Table 3.2) and proceed to the relevant section(s) below for confirmatory tests or for more detail of the diagnostic method.

3.3.2.1 **Group Ia (LOPAT + − − − +) and Ib (LOPAT − − − − +) fluorescent pseudomonads**

A general method for isolation and species confirmation in the group follows. Symptoms and additional tests largely based on host tests are given below for specific, named diseases.

Isolation and test procedures. Streak suitable suspensions from diseased material (1.3 and 1.4) on:
1 5% sucrose nutrient agar (6.2.17).
2 King's medium B (6.2.7).

Incubate for 2–3 days. If levan colonies are present on SNA, inoculate one set of each of the following media from each of 2–3 water suspensions of well-isolated colonies:
1 Nutrient agar slope (6.2.9).
2 King's medium B plate (6.2.7).
3 Potato slice (5.3.6) or pectate plate (6.2.12).
4 Arginine medium (6.4.3).

After 2 days examine the King's medium B plates. If any are obviously mixtures of bacteria, ignore the set of LOPAT test media made from the same colonies. For apparently pure cultures, test the growth from the nutrient agar slopes for oxidase reaction (6.6.2) and use water suspensions of the growth to inoculate tobacco plants (5.2.1). After 2 days, test potato slices (1 day and 2 day if pectate plates are used) for pectolytic activity. Record fluorescence on re-streaked King's medium B plates. At 3 days record the presence or absence of arginine dihydrolase. Again examine the re-streaked King's medium B plates and view with suspicion any set of LOPAT tests corresponding to a plate with mixed colony types.

If the 5% sucrose nutrient agar plates are overgrown with contaminants, pick green-fluorescent colonies from King's medium B plates and streak them on 5% sucrose nutrient agar plates. Then proceed to the tests given above.

Presumptive diagnosis. A positive presumptive diagnosis of a disease caused by a group Ia pathogen can be given if green-fluorescent cultures which produce the reactions given for the group in Table 3.2 are isolated from plants with typical symptoms.

It should be noted that a negative tobacco reaction is, so far as is known, certain evidence that an isolate is not a group Ia pathogen. *P. syringae* pv. *tabaci*, of course, produces a pathogenic but not a hypersensitivity reaction on tobacco.

It is also important to note that occasionally atypical non-fluorescent strains of normally green-fluorescent pathovars of group I are isolated; that in some pathovars, pv. *morsprunorum* is one such, non-fluorescent isolates on King's medium B are common; and that in pv. *persicae* all isolates are non-fluorescent on this medium so far as is known. On other media (6.2.7 and 6.2.8) such isolates often produce the pigments. Certain group I members do not produce levan, are separated into group Ib and include most *P. s.* subsp. *savastanoi* and all tested *P. s.* pv. *delphinii* isolates. Some isolates of *P. s.* subsp. *savastanoi* produce mucoid colonies not dissimilar to levan-type colonies. Colonies of *P. s.* pv. *delphinii* are typically butyrous.

Confirmation of diagnosis. The additional tests in Table 3.2 can be performed to confirm the diagnosis to species level. For further information see Lelliott *et al.* (1966).

Diagnosis to pathovar level can only be made by a host test and is then not always possible. If suitable test plants are readily to hand, confirmation at species level is usually better and more quickly obtained by a host test than the laboratory tests given above; for details of host tests see Chapter 5, but the following general points need to be made here. A range of hosts including the host from which the pathogen was isolated is desirable. Inocula should not contain large numbers of bacteria. Populations of 10^6 to 10^7 colony-forming units (cfu)/ml are generally best. Plants should be young and for certain diseases a seedling test is useful. When young plants are not available, young shoots or fruits should be inoculated by an appropriate method.

*Lilac blight, pear blossom blight (*Pseudomonas syringae *pv.* syringae*)*

Symptoms. Many blight symptoms on these hosts are probably the result of ice nucleation activity by *P. syringae* pv. *syringae*. On lilac (*Syringa* spp.) a general blight is often found in which small, brown-black angular spots occur on leaves often causing distortion of the lamina; longitudinal black spots on stems may often girdle and kill them; dark spots may occur on buds and flowers which may coalesce preventing bud burst or even killing the whole inflorescence.

On Pear (*Pyrus* spp.) the symptoms can be easily confused with Fire-blight caused by *Erwinia amylovora*. Shoot dieback is perhaps the most common symptom. Leaves develop rapidly-expanding necrotic spots which may kill whole leaves and progress rapidly down stems. Blossom infections and blight is common, and if infection progresses from spurs into branches, sunken cankers occur, the internal tissue of which is dark-green to brown.

Many strains will produce reddish-brown lesions on dwarf bean pods (5.3.2.2). Strains from lilac can be tested on young, rapidly-growing leaves and stems of detached lilac shoots (5.3.4), dormancy of which may be broken, if necessary, under artificial light (12 hours–16 hours per day). Strains from fruit trees can be tested on pear fruitlets. Diagnosis of pear blossom blight is considered in detail in 3.4.3.

*Leaf spot, shot hole and bacterial canker of cherry and plum (*Pseudomonas syringae *pv.* morsprunorum*) See also 3.4.4*

This disease is common in Europe. The main hosts are cherry and plum. Strains are specific for each host. Cherry strains are cultivar specific and at least two races are recognized. The disease is virtually indistinguishable from a similar disease caused by *P. syringae* pv. *syringae* which is equally common in some areas of mid-Europe.

Symptoms. Spots on leaves are usually reddish–dark brown, rounded or angular, often coalescing to form large, irregular, necrotic spots which may drop out to give a shot hole effect. Stem symptoms vary with host. On cherry, cankers develop usually in the crotch, the angles between branches, and around leaf scars and fruiting spurs. When cankers girdle the stem, dieback occurs. On plum, cankers are usually dark, long, strap-like depressions in the bark. Gummy exudates are found on both hosts, but are much more common in cherry infections. For further information on symptoms and disease biology see 3.4.4 and Crosse (1966), Crosse and Garrett (1966), Garrett *et al.* (1966), and Garrett (1986).

The pathogen dies out or is present in very low numbers in leaf spots in the summer, but can be recovered from surface washing of leaves. It can be distinguished from *P. syringae* pv. *syringae* by the characters shown in Table 3.4.

Table 3.4. Characters distinguishing *Pseudomonas syringae* pv. *morsprunorum* from *P. syringae* pv. *syringae*

Character	pv. *morsprunorum*	pv. *syringae*
5% sucrose nutrient broth (6.2.17)	White growth	Yellow growth
Recovery from 5% sucrose nutrient agar after 6 days (6.2.17)*	−	+
Aesculin or arbutin hydrolysis (6.5.4)	−	+
Gelatin liquefaction (6.5.2)	−	+
Brown diffusible pigment on King's medium B (6.2.7)	±	−
Absence of fluorescence on King's medium B (6.2.7)†	±	−

*Test by re-streaking on a nutrient agar plate.
†Colonies of pv. morsprunorum on King's medium B usually do not produce a diffusible, green-fluorescent pigment but may produce a brown diffusible pigment. Colonies of pv. syringae usually produce such a pigment.

*Leaf blight of wheat (*Pseudomonas syringae *pv.* syringae*)*

Symptoms. Small, watersoaked lesions, which later enlarge, become pale and necrotic and may coalesce to form irregular blotches.

Miscellaneous diseases attributed to Pseudomonas syringae *pv.* syringae

A wide variety of other leaf spot and necrotic diseases are attributed to *P. s.* pv. *syringae.* These include a wide range of cereals and grasses (e.g. maize, sorghum), legumes (pea, various genera of bean), woody ornamental or fruit tree genera (*Hibiscus, Philadelphus, Malus, Rosa, Juglans*). Some of these are almost certainly host specific pathovars.

*Basal glume rot of wheat (*Pseudomonas syringae *pv.* atrofaciens*)*

This is a common disease in many wheat-growing areas of the world, especially where rainfall is high just after flowering. Not usually a commercially important disease. The bacterium is known to infect several grasses as well as wheat.

Symptoms. Dark-green, watersoaked streaks occur uncommonly on leaves and dark longitudinal stripes on culms. Glumes are streaked dark and usually discoloured only at the bases. Kernels may be discoloured, dark to black, and are often shrunken. See Wilkie (1973).

*Halo blight of beans (*Pseudomonas syringae *pv.* phaseolicola*)*

Symptoms. This disease derives its name from the characteristic leaf spots. These are semi-transparent to dark-brown and watersoaked, up to about 3 mm in diameter, each one being surrounded by a yellow halo. If spots are numerous or infection occurs during hot weather these haloes may be absent and the disease is then very difficult to differentiate on symptoms from that caused by *Xanthomonas campestris* pv. *phaseoli.* Spots may coalesce and dry up, causing large necrotic areas. Pod infections appear as circular, watersoaked spots from which masses of bacteria ooze. These may look like drops of candle grease and later be surrounded by a reddish-brown margin.

Systemic symptoms are seen mainly in young plants and may be difficult to diagnose. Leaves often show a chlorotic mosaic pattern reminiscent of a viral disease. In severe epidemics where stems are badly infected, the leaves or even the whole plant may wilt, and leaf veins, petioles and the stem vascular system may show a reddish discoloration and bacterial ooze, often in 'tendrils'. Similar leaf spots have been reported in the USA caused by *P. syringae* pv. *syringae* strains, but these typically produce ring spots on pods. For further details of symptoms see Zaumeyer (1932), Hayward and Waterston (1965), Taylor *et al.* (1979).

The following tests may be used to separate the bacteria.

	pv. *phaseolicola*	pv. *syringae*
aesculin or arbutin hydrolysis (6.5.4)	−	+
gelatin liquefaction (6.5.2) within 3 days	−	+
'grease spots' on detached bean pods (5.3.2.3)	+	±
utilization of sorbitol, inositol and erythritol (6.3.2)	−	+

As with the majority of such schemes, isolates with atypical characters are likely to be found.

*Bacterial blight of soybean (*Pseudomonas syringae *pv.* glycinea*)*

This bacterium causes a serious disease of soybean, especially in the USA.

Symptoms. On leaves small, watersoaked angular spots initially appear yellow, turn light brown, becoming brown and necrotic, usually surrounded by a yellow halo. As lesions enlarge, they may coalesce and disintegrate, giving a tattered effect. Occasional large, dark, brown-black lesions may occur on stems. Similar spots may develop on pods. Infected seeds may also show lesions and may exude bacteria-laden ooze.

Isolation and species identification are as for group I pseudomonads (3.3.2.1). The disease can be confirmed by a rapid host test using seedlings with fully-developed bean cotyledons as for *P.s.* pv. *glycinea* (5.3.4.4).

*Bacterial blight of peas (*Pseudomonas syringae *pv.* pisi*)*

Symptoms. All aerial parts of the plant may be affected. Spots on leaves, stipules, stems, tendrils and pods are watersoaked, in many cases turning pale brown often with darker margins. Characteristic fan-shaped lesions may be found on stipules and leaflets. Infected pods and leaves may shrivel and seeds may be discoloured or even covered with bacterial ooze. Heavy infections may become systemic. For further details of symptoms see Cirulli and Ercolani (1969), Taylor (1972), Taylor and Dye (1972).

P. s. pv. *syringae* can cause a very similar disease early (as a result of ice nucleation damage) or sometimes late in the season. This may often be more damaging than that caused by *P. s.* pv. *pisi*. No biochemical tests are known which can distinguish between the two pathovars with acceptable accuracy, although *P. s.* pv. *pisi* is reputed to be one of the few pathovars able to utilize homoserine, but some *P. s.* pv. *syringae* strains from pea may also do this. Serological tests cannot always be reliably used to differentiate between the two pathovars. Certain phages specific for *P. s.* pv. *pisi* are known. Rapid production of watersoaked lesions after a stem stab in young pea stems is the best method for diagnosis of *P. s.* pv. *pisi* (5.3.5.1). Five races of *P. s* pv. *pisi* are known. Pea cultivar Kelvedon Wonder is susceptible to all five races.

*Angular leaf spot of cucumber and other cucurbits (*Pseudomonas syringae *pv.* lachrymans*)*

Symptoms. The most characteristic symptom is the tear drop produced from leaf and fruit spots in humid conditions. As these drops dry out leaves and fruit may be covered with papery, thin, dried exudate. Leaf spots are usually yellowish-brown and angular. Leaves may have a tattered appearance and fruits may fall prematurely. For further details of symptoms see Bradbury (1967b), Pohronezny *et al.* (1978).

Isolation is best from active spots producing tear drops, or from the drops themselves. Diagnose as for any group I fluorescent pseudomonad using the LOPAT scheme (Table 3.2 and 3.3.2). For host test see 5.3.4. Stab inoculation of cucumber fruits should reproduce the tear drop symptom (5.3.2.4).

*Pepper spot, leaf spot of brassicas (*Pseudomonas syringae *pv.* maculicola*)*

Symptoms. Infected stomata develop into typical but small, watersoaked, brown-purple spots giving a flecked appearance to the diseased leaves, petioles and stems. Other pseudomonads, e.g. *P. viridiflava* may cause leaf spots but these are usually larger. For further details of symptoms see Goldsworthy (1926), Smith and Ramsey (1953), Williams and Keen (1966).

Isolate and identify as for group Ia pseudomonads. *P. syringae* pv. *syringae* may also cause similar symptoms, sometimes difficult to distinguish from pv. *maculicola*. Symptom expression is an important diagnostic feature. For host test see 5.3.4.

*Bacterial speck of tomato (*Pseudomonas syringae *pv.* tomato*)*

Symptoms. This bacterium produces small, brown spots with yellow margins on tomato leaves. In outdoor crops these are often more common along the margins of the leaves. Dark, raised, scab-like specks and spots develop on fruits. *P. syringae* pv. *syringae* may also cause similar symptoms. For further details of symptoms see Wilkie and Dye (1974b).

Isolate as for a group Ia fluorescent pseudomonad. *P. syringae* pv. *tomato* can be distinguished from pv. *syringae* by several nutritional tests.

Inoculate media for utilization of erythritol and L-lactate (6.3.2). Both are used as sole carbon and energy sources by pv. *syringae* but not pv. *tomato*. For host test see 5.3.4 and 5.3.4.3.

*Halo blight of oats (*Pseudomonas syringae *pv.* coronafaciens*)*

Symptoms. This bacterium causes oval, light-green to pale-brown spots, most numerous along the margins, with watersoaked margins and characteristic yellow haloes up to 1 cm diameter. Spots are centred on stomata and can occur on glumes. Spikelets often fail to develop. Other cereals and grasses are susceptible but the disease is most serious in oats. Further details of symptoms may be obtained from Wilkie (1972).

Isolate and diagnose as for group Ia fluorescent pseudomonads. For host test see 5.3.4.

*Antirrhinum blight (*Pseudomonas syringae *pv.* antirrhini*)*

Symptoms. This disease is usually most serious as a seedling blight, producing watersoaked leaf and stem spots which coalesce. Seedlings and young plants may die.

Isolate and diagnose as for a group Ia fluorescent pseudomonad. For host test see 5.3.4.

*Tobacco wildfire (*Pseudomonas syringae *pv.* tabaci*)*

Symptoms. Leaf and stem spots are generally reddish-brown to brown-black in colour and have chlorotic, yellow haloes. As they enlarge they become limited by leaf veins and appear angular and may even fall out. Some non-toxin producing strains previously classified as *Pseudomonas angulata* produce spots with no haloes. *P. syringae* pv. *tabaci* may also cause a halo blight of soybean.

Isolate and perform LOPAT tests as for group Ia fluorescent pseudomonads. The bacterium can be distinguished with practice from *P. s.* pv. *syringae* by its pathogenic instead of hypersensitivity reaction on tobacco which is characterized by a more slowly-developing water-soaking of the inoculated tissues that do not rapidly become dry, light-coloured and papery but dark and necrotic. For host test see 5.3.4. For further information see Bradbury (1967c).

*Black blotch of delphinium (*Pseudomonas syringae *pv.* delphinii*)*

Symptoms. Irregular, dark, brown-black blotches appear on leaves, flowers and stems. On leaves they are raised on the upper surface and sunken on the lower.

Isolate and diagnose as for group Ib pseudomonads (3.3.2.1). Inoculate King's medium B (6.2.7) and 5% sucrose nutrient agar (6.2.17). Look for levan-negative, fluorescent colonies. Purify and recheck on both media, comparing growth on the 5% sucrose nutrient agar plate with that on a nutrient agar plate. *P. syringae* pv. *delphinii* does not produce levan demonstrated by similar, non-mucoid growth on nutrient and 5% sucrose nutrient agars, typical of group Ib. For host test see 5.3.4.

*Mulberry blight (*Pseudomonas syringae *pv.* mori*)*

This is a common and widespread disease of mulberry.

Symptoms. Leaf spots are small and brown-black, usually with yellow haloes. Infected leaves are often distorted. Infected buds may become disfigured as they swell. Young shoots may show rapid necrosis and dieback. Occasional stem cankers occur which may exude bacteria-laden ooze. Infected bushes and trees often appear stunted. Further information on symptoms can be obtained from Doidge (1915) and Wormald (1939).

Isolate and diagnose as for group Ia fluorescent pseudomonads. For host test see 5.3.3 and 5.3.4.

3.3.2.2 Necrotic diseases caused by group II (LOPAT − − + − +) fluorescent pseudomonads (*Pseudomonas viridiflava*)

This is an opportunist pathogen and is widely distributed on plant surfaces. Under suitable conditions it can cause spots and general necrosis of a wide variety of herbaceous dicotyledonous plants including bean, cabbage, chrysanthemum, parsley, tomato, primula and lettuce. It is often isolated as a contaminant in lesions caused by other pathogens including *Erwinia amylovora* and pathovars of *Pseudomonas syringae.*

Isolation. Streak suitable suspensions from diseased material (1.3 and 1.4) on 5% sucrose nutrient agar (6.2.17). Incubate for 2–3 days and look for non-levan type (6.3.9), whitish to pale-yellow colonies with green centres. Long incubation (up to 7 days) may be required for production of green centres. The green centre is a very useful diagnostic character on isolation plates but is easily lost in subsequent subculture.

Test procedures. From each of 2–3 water suspensions of typical, well-isolated colonies inoculate a set of the following LOPAT media, streak plates of King's medium B (6.2.7) and store the 5% sucrose nutrient agar isolation plates at 2–4°C:

1 Nutrient agar slope (6.2.9).
2 Potato slice (5.3.6) or pectate plate (6.2.12).
3 Arginine medium (6.4.3).

If no typical, green-centred colonies are present on the isolation plate proceed similarly from any non-pigmented colonies that have the other colony characters of *P. viridiflava* and that have been shown by Gram stain to be composed wholly of Gram-negative cells.

Incubate for 2 days and if the re-streaked King's medium B plates are obviously a mixture, ignore the LOPAT set made from the same suspension. For LOPAT set(s) from suspensions shown by the re-streaking to be apparently pure, test for oxidase reaction (6.6.2) on growth from the nutrient agar slope culture and fluorescence on King's medium B. Suspend the remainder of the nutrient agar slope in sterile water and test for tobacco hypersensitivity (5.2.1). Record pectolytic activity. After 3 days record arginine dihydrolase activity. Again examine the re-streaked King's medium B and regard with suspicion any set of LOPAT results inoculated from the same suspension as plates with different colony types. If no such plates are of a single, green-fluorescent colony type re-streak green-centred colonies from the original 5% sucrose nutrient agar isolation plate and recommence the test procedure.

Presumptive diagnosis. A presumptive diagnosis of necrosis by *P. viridiflava* can be given if green-fluorescent isolates are obtained with the colony character given above and the LOPAT characteristics given in Table 3.2 from plants with the characteristic symptoms.

Confirmation of diagnosis. Confirm a diagnosis of *P. viridiflava* if green-fluorescent cultures which are isolated also have the secondary LOPAT characteristics of group II (− − − −) given in Table 3.2 and if a positive host test is obtained. The spray technique (5.3.4) on a variety of hosts, including that from which isolation was made, is normally the best method. Cultures of *P. viridiflava* can lose pathogenicity within a few months in culture. Further information on the pathogen can be obtained from Billing (1970) and Lelliott *et al.* (1966). Symptoms and host ranges are described by Wilkie *et al.* (1973) and Jones *et al.* (1984).

3.3.2.3 Leaf and stem necroses caused by group III (LOPAT − + − − +) fluorescent pseudomonads (*Pseudomonas cichorii*)

These are opportunist pathogens and may infect a wide variety of herbaceous dicotyledonous plants, especially lettuce, chicory, onion, chrysanthemum, begonia and dahlia.

Symptoms. Black, usually rapidly-spreading and often necrotic spots with watersoaked margins develop on leaves and spots with watersoaked margins on stems. An unusual symptom, varnish spot, has been recorded on lettuce (Grogan *et al.*, 1977).

Isolation. Streak suitable suspensions from plant material on King's medium B (6.2.7). Incubate for 2 days.

Test procedures. Suspend each of 2–3 well-separated, fluorescent colonies from the isolation plates in sterile water, and from each suspension streak another King's medium B plate and inoculate a set of the following media:

1 Nutrient agar slope (6.2.9).
2 Potato slice (5.3.6) or pectate plate (6.2.12).

3 Arginine medium (6.4.3).
4 5% sucrose nutrient agar plate (6.2.17).
Store the isolation plates at 2–4°C.

Incubate and after 2 days examine the re-streaked King's medium B plates. If any have an obvious mixture of bacteria ignore the results from the set of LOPAT tests inoculated from the same suspension. For any remaining sets, test growth from the nutrient agar slope for oxidase reaction (6.6.2). The oxidase reaction of group III is weaker than group IV and it is essential to use a heavily-charged loop. Use the remainder of the growth for a tobacco hypersensitivity test (5.2.1). Test for pectolytic activity.

If none of the re-steaked plates are apparently pure cultures of green-fluorescent bacteria, pick a selection of green-fluorescent colonies and from these inoculate fresh sets of LOPAT media and King's medium B plates. After 3 days look for levan colonies on 5% SNA and record the arginine reaction.

Re-examine the re-streaked King's medium B plates and if any are of more than one colony type treat the corresponding LOPAT test results with suspicion.

Presumptive diagnosis. A presumptive diagnosis of a disease caused by a group III pseudomonad can be made if green-fluorescent cultures with the LOPAT characteristics of group III (Table 3.2) are isolated from plants with the characteristic symptoms.

Confirmation of diagnosis. Diagnosis can be confirmed if the results of LOPAT confirmatory tests have the characteristics (– – d –) given in Table 3.2 and host tests on young or seedling lettuce or chicory are positive using spray inoculations (5.3.4).

For further information on the disease see Wilkie and Dye (1974a) and Grogan *et al.* (1977).

3.3.2.4 Necrotic diseases caused by group IV (LOPAT ± + + + –) fluorescent pseudomonads

Taxonomically there are good grounds for regarding the pathogens in this group as pectinase-producing members of *Pseudomonas fluorescens* or *P. putida* (Palleroni 1984). However, it is not certain that all pectinase-producing members of these two species (most of whose members are saprophytes) are plant pathogens or in which of the two species the pathogens of group IV should be placed. Those isolates studied and commonly included in *P. marginalis* are all members of the levan-positive group IVa and are more closely related to *P. fluorescens* than to other plant pathogens of the genus. It seems preferable to retain the name *P. marginalis* (which is legitimate) for the pathogens of group IVa and to refer to those of the levan-negative group IVb as pathogenic strains of the fluorescens/putida complex of the genus *Pseudomonas*. *P. marginalis* has been subdivided into three pathovars, pv. *marginalis*, pv. *alfalfae* and pv. *pastinaceae* (Young *et al*, 1978) but the validity for this is doubtful.

Symptoms. P. marginalis causes an economically important foliar disease characterized by marginal necrosis and spotting of leaves often associated with black discoloration of the leaf veins. It is an opportunist pathogen, as are the members of group IVb, causing soft rotting and a wide range of necrotic symptoms on many hosts. In lettuce and some other soft-leaved plants, marginal spots occur with darkening of the veins; generally lesions are watersoaked and rapidly turn dark-brown or black. On stems, lesions are watersoaked and become brown and necrotic. The levan-negative members of IV (i.e. group IVb) have been less well studied and their symptomatology is not clear.

Isolation. Streak suitable suspensions from plant material (1.3 and 1.4) on 5% sucrose nutrient agar (6.2.17) and King's medium B (6.2.7). Incubate for 2–3 days and look for well-separated levan-type colonies (6.3.9) from the 5% sucrose nutrient agar plate. If levan colonies are not present, streak representative green-fluorescent colonies (if present) from the King's medium B plate onto pectate gel (6.2.12).

Test procedures. Suspend each of 2–3 well-separated levan-type colonies or if absent, pectate-degrading colonies in sterile water and from each suspension streak a plate of King's medium B and inoculate a set of the following media:

1 Nutrient agar slope (6.2.9).
2 Arginine medium (6.4.3).
3 Potato slice (5.3.6) or pectate plate (6.2.12).

Disregard sets inoculated from suspensions that give mixed colony forms on the King's medium B plate.

After 2 days read the oxidase reaction from the nutrient agar slope and use the remainder of the culture for a tobacco hypersensitivity test (5.2.1). Record pectinase activity. After 3 days observe the arginine dihydrolase activity.

If the 5% sucrose nutrient agar isolation plate is overgrown, suspend typical, fluorescent colonies from the King's medium B plate in sterile water and proceed as above.

Presumptive diagnosis. A presumptive diagnosis of a disease caused by

P. marginalis can be made if green-fluorescent colonies are isolated with the primary characteristics of group IVa, as defined by the LOPAT tests in Table 3.2 from plants with characteristic symptoms. If such colonies have the primary characteristics of group IVb diagnose pathogenic strains of the fluorescens/putida complex of *Pseudomonas*.

Note that strains of *P. marginalis* from parsnip, referred to pv. *pastinaceae*, produce pink, diffusible pigment on 5% sucrose nutrient agar.

Confirmation of diagnosis. Generally speaking, the pathogens of this group are so easily differentiated from other pseudomonad pathogens that testing for the subsidiary LOPAT characters is unnecessary. The main difficulty in diagnosis of this group is to differentiate the situation where the group IV pathogen isolated is the primary pathogen from that when it is secondary to another pathogen. A host test on plants of the host from which isolates were obtained and careful consideration of the symptoms will usually resolve this difficulty.

A presumptive diagnosis can therefore be confirmed if isolates can reproduce the symptoms seen in the original specimen in a host test. Host range and symptoms are discussed by Brocklehurst and Lund (1981).

3.3.3 Non-fluorescent pseudomonads

With few exceptions (mainly pv. *persicae* and many strains of pv *pisi* and pv. *morsprunorum* of *P. syringae*) plant pathogenic pseudomonads that do not fluoresce on King's medium B (6.2.7 and 6.2.8) are members of the section of the genus that produces poly-β-hydroxybutyrate (PHB) as a carbon reserve material. All pathogenic members of that section except *P. solanacearum*, *P. caryophylli* and *P. cepacia* cause spots and necroses of leaves, stems and fruits.

3.3.3.1 Leaf spots of monocotyledons caused by *Pseudomonas avenae*

Formerly known as *P. alboprecipitans*, this bacterium is very closely related to and may be synonymous with several other species including *P. setariae* and *P. panici*, none of which are included in the approved lists (Skerman *et al.*, 1980).

Symptoms. P. avenae causes bacterial leaf blight and stalk rot of maize, leaf blight of oats, leaf spot of sorghum and bacterial brown stripe of many grass species. Spots, streaks or stripes occur on leaves, sheaths and culms and are usually brown or reddish. Stems may occasionally rot.

Isolation. Streak suitable suspensions from plant material (1.3 and 1.4) on nutrient agar plates (6.2.9). *P. avenae* produces greyish-white, smooth, entire colonies, often producing an opalescent haze (hence the synonym albo-precipitans) around the colonies. Gram stain likely colonies. Suspend each of 2–3 such colonies that are Gram-negative in sterile water and from each suspension inoculate a set of the following media:

1 King's medium B slope (6.2.7).
2 Nutrient gelatin (6.5.2).
3 Ayers Rupp & Johnsons medium (6.3.2) + 0.5% β-hydroxybutyrate.
4 Starch plate (6.5.3).
5 O/F medium (6.3.1).
6 Nitrate reductase broth (6.4.1).

Also streak a nutrient agar plate.

After 2–3 days check the nutrient agar plates and disregard sets of media inoculated from the same suspension as any plate that has mixed colony forms. Use growth from the nutrient agar plates to determine catalase (6.6.1) and oxidase activity (6.6.2). Record the O/F reaction and test King's medium B for fluorescence production. After 3–4 days test for poly-β-hydroxybutyrate production (6.1.5), gelatin liquefaction and hydrolysis of starch.

Presumptive diagnosis. A presumptive diagnosis of a leaf-spot disease caused by *P. avenae* can be made if isolates from plants with typical symptoms have the colony morphology of this organism, are Gram-negative, oxidative, catalase and oxidase-positive, reduce nitrate to nitrite, do not liquefy gelatin, rapidly hydrolyze starch, do not produce fluorescent pigments and produce poly-β-hydroxybutyrate.

Confirmation of diagnosis. Confirm the diagnosis if a host test (5.3.4) reproduces the symptoms on the original specimen. For further information on the disease see Bradbury (1973a), Schaad *et al.* (1975).

3.3.3.2 Leaf spots caused by *Pseudomonas andropogonis*

P. andropogonis is a heterogeneous species causing leaf spots and blights of a variety of grasses and cereals (including sorghum), legumes, carnation and *Bougainvillea.* Some isolates are host specific; others have a fairly broad host spectrum. Two pathovars have been proposed: pv. *andropogonis*, the agent of stripe disease of sorghum and pv. *stizolobii*, the cause of leaf spot of velvet bean (Fuerst & Hayward, 1980).

Symptoms. On sorghum, spots and stripes occur on leaves and sheaths and are usually red to purple in colour. On legumes they tend to be smaller, more angular and reddish-brown, dark brown or black with yellow haloes. Nishiyama *et al.* (1979) regard *P. woodsii* as synonymous with *P. andropogonis.*

Isolation. Streak suitable suspensions from plant material (1.3 and 1.4) on nutrient agar and look for greyish-white, viscid or butyrous colonies which are smooth and entire.

Test procedure. Gram stain any likely colonies, and from each of 2–3 such colonies that are Gram-negative prepare separate suspensions in sterile water. From each suspension inoculate a set of the following media:

1 King's medium B slope (6.2.7).
2 O/F medium (6.3.1).
3 Nutrient gelatin (6.5.2).
4 Ayers Rupp & Johnson's medium (6.3.2) + 0.5% β-hydroxybutyrate.
5 Nitrate reductase broth (6.4.1).
6 Arginine dihydrolase medium (6.4.3).

Also streak a nutrient agar plate (6.2.9).

Disregard any sets of media in which the nutrient agar plate appears to be a mixed culture. Otherwise, after 2–3 days use colonies from nutrient agar to determine catalase (6.6.1) and oxidase reaction (6.6.2). Record fluorescence on King's medium B, whether oxidative or fermentative on O/F medium, gelatinase activity, production of poly-β-hydroxybutyrate (6.1.5), reduction of nitrate and arginine dihydrolase activity.

Presumptive diagnosis. A presumptive diagnosis can be made if isolates from plants with typical symptoms are Gram-negative, oxidative, oxidase-negative, catalase-positive, arginine dihydrolase-negative, do not reduce nitrate, do not produce fluorescent pigment or liquefy gelatin, are butyrous becoming viscid on nutrient agar and produce poly-β-hydroxybutyrate.

Confirmation of diagnosis. A diagnosis of leaf spot caused by *P. andropogonis* can be confirmed if isolates cause typical symptoms when inoculated into plants of the host from which they came. If such plants are not available, inoculate young plants of bean (*Phaseolus vulgaris*), clover (*Trifolium repens*), sorghum and maize (5.3.4). Most isolates will infect at least one of these hosts. On bean and clover, symptoms appear as small, dark leaf spots; on sorghum and maize, as reddish-brown leaf streaks. For further information see Goto and Starr (1971), Robbs *et al.* (1981).

3.3.3.3 Red stripe of sugar-cane (*Pseudomonas rubrilineans*)

P. rubrilineans causes red stripe of sugar-cane, a disease found in most sugar-cane growing areas of the world. As well as the classic, long, dark-red stripes, the apical shoot may become necrotic, a condition often referred to as 'top rot'. This phase of the disease is more damaging to the crop and may occur in the absence of red stripe.

Symptoms. Leaf stripes are usually watersoaked and have chlorotic margins which become reddish-brown to dark-red as they age. The stripes often extend down the leaves to the bases of the sheaths. Vascular bundles often show a red discoloration, often in a ring. The central tissue may rot, leaving a hollow stem. Terminal buds may die and rot, hence the term top rot.

Isolation. Streak suitable suspensions of plant material (1.3 and 1.4) on sucrose peptone agar (6.2.18.1) and nutrient dextrose agar (6.2.10). Incubate and look for cream-coloured, fimbriate or crenate, low-convex or raised, butyrous colonies. Gram stain (6.1.3) such colonies and suspend each of 2–3 colonies that are negative separately in sterile water and from each inoculate a set of the following media:
1 O/F medium (6.3.1).
2 Nitrate reductase broth (6.4.1).
3 Christensen's urease medium (6.4.2).
4 Ayers, Rupp & Johnson's medium (6.3.2) + 0.5% β-hydroxybutyrate.
5 Aesculin medium (6.5.4).
6 Tween 80 medium (6.5.5).
7 Tryptone broth for indole production (6.4.4).
8 Gelatin (6.5.2).
9 King's medium B tube (6.2.7).
Stain for the presence of poly-β-hydroxybutyrate granules (6.1.5) and streak on a nutrient agar plate (6.2.9). Incubate for 3 days. Disregard any sets of media in which the growth on the nutrient agar plate appears to be a mixed culture. Otherwise record whether oxidative or fermentative in O/F medium and oxidase reaction from the nutrient agar plate; record reduction of nitrate, urease activity, hydrolysis of aesculin and Tween 80, production of indole, gelatinase activity and fluorescence on King's B medium. Test for poly-β-hydroxybutyrate (6.1.5).

Presumptive diagnosis. A presumptive diagnosis of red stripe can be made if bacteria are isolated that are oxidative, oxidase positive, produce urease, and hydrolyze Tween 80 but do not liquefy gelatin or produce indole, do not hydrolyze aesculin but reduce nitrate to nitrite, do not produce fluorescent pigments, and accumulate poly-β-hydroxybutyrate (6.1.5).

Confirmatory diagnosis. Members of the non-fluorescent group of pseudomonads can be difficult to distinguish. Confirm a diagnosis of *P. rubrilineans*, which is very closely related to *P. rubrisubalbicans*, if typical symptoms of the disease are reproduced on young leaves by a host test (5.3.4). Further information can be obtained from Martin and Wismer (1961).

3.3.3.4 Mottled stripe of sugar-cane (*Pseudomonas rubrisubalbicans*)

P. rubrisubalbicans causes a widespread disease of sugar-cane known as mottled stripe. Although common in many sugar-cane producing areas it is probably not so widespread as red stripe and is considered to be less damaging to the crop.

Symptoms. Stripes are a pale-red colour and are distinctively speckled with white and have irregular white margins. They are more commonly found on the lower parts of the leaves.

Isolation. Streak suitable suspensions of plant material (1.3 and 1.4) on sucrose peptone agar (6.2.18.1). Incubate and examine for small, whitish-to-buff, raised, erose colonies. Make Gram strains from any such colonies and a suspension from each of 2–3 that are Gram-negative. Streak a sucrose peptone plate and inoculate a set of the following media from each suspension:

1 O/F medium (6.3.1).
2 Nitrate reductase broth (6.4.1).
3 Christensen's urease medium (6.4.2).
4 Tween 80 medium (6.5.5.).
5 Tryptone broth for indole production (6.4.4).
6 Ayers, Rupp & Johnson's medium (6.3.2) + 0.5% β-hydroxybutyrate.
7 Gelatin (6.5.2).
8 King's medium B (6.2.7).

Stain for the presence of poly-β-hydroxybutyrate granules (6.1.5). Incubate until isolated colonies are apparent on the sucrose peptone plates and then test colonies for oxidase reaction. Determine the results on the other test media. Disregard the results of any set of tests derived from suspensions giving mixed colony forms on the sucrose peptone plates.

Presumptive diagnosis. A presumptive diagnosis of mottle stripe can be made if bacteria with the following characteristics are isolated from tissue affected with typical symptoms of the disease:

1 Gram reaction negative.
2 Oxidase reaction negative.
3 Oxidative.
4 Green fluorescent pigment produced.
5 PHB produced.
6 Nitrate reduced to nitrite.
7 Urease positive.
8 Indole produced.
9 Gelatin not liquefied.
10 Tween 80 not hydrolyzed.

Confirmation of diagnosis. The diagnosis can be confirmed if typical symptoms are produced on sugar-cane by a host test (5.3.4). For further information see Christopher and Edgerton (1930).

3.3.3.5 Tomato pith necrosis; stem bacteriosis (*Pseudomonas corrugata* or *P. cichorii* or *P. viridiflava*)

In England and probably much of western Europe this disease can reasonably be attributed to *P. corrugata*. However, a similar disease elsewhere has been attributed to other pseudomonads, including *P. viridiflava* (Wilkie *et al.* 1973) and an as yet unidentified member of LOPAT group Ia. Particularly, the description of a disease in New Zealand, stem bacteriosis ascribed by Wilkie & Dye (1974a) to *P. cichorii*, closely resembles that of pith necrosis. This latter organism has been looked for but not found in many cases of pith necrosis in England but diagnosticians should always be aware of the possibility that either organism is associated with the symptoms below and if the primary isolation plates appear to yield green-fluorescent pseudomonads, to follow up this possibility by using the diagnostic procedure described for *P. cichorii* (3.3.2.3), particularly if they do not appear to yield *P. corrugata*. The following scheme is intended to detect *P. corrugata*.

Symptoms. This disease affects mature tomato plants. The main symptom is a brownish discolouration and/or necrosis and collapse of the pith, sometimes accompanied by wilting, by vascular browning, by dark-brown lesions on the stem surface and by whitish bacterial exudate from wounds. The pith browning usually extends far up the plant from the base and in to the peduncles. As it disintegrates it produces a characteristic collapse and 'laddering' of the pith.

Isolation. Make suspensions from the edge of necrotic lesions (1.4) or from ooze, examine microscopically (1.3) and plate those containing bacteria on to nutrient agar, King's medium B (6.2.7) and nutrient dextrose agar (6.2.10). After 2 days on nutrient dextrose agar look for yellowish colonies, often with blue-green centres, which often produce a yellow-green diffusible pigment and which may be wrinkled; wrinkling becomes more apparent after prolonged incubation. On King's medium B *P. corrugata* colonies produce a diffusible yellow-green pigment which does *not* fluoresce in ultra-violet light.

On nutrient agar after two days look for smooth colonies with a pale golden-brown pigment; this pigment is more pronounced in the centre of the colony, and usually intensifies with prolonged incubation. Consult Scarlett *et al.* (1978) for more detailed colony description.

Test procedures. Make Gram stains from a range of likely-looking, well-isolated colonies. Prepare a separate suspension from each of 2–3 such colonies that contain only Gram-negative bacterial cells. From each suspension re-streak a plate of nutrient agar (6.2.9) and inoculate a set of the following media:

1 O/F medium (6.3.1).
2 Egg yolk medium (6.5.1).
3 Nitrate reductase medium (6.4.1).
4 Arginine dihydrolase medium (6.4.3).
5 Nutrient agar slope (6.2.9).
6 Ayers, Rupp & Johnson's medium (6.3.2) + 0.5% β-hydroxybutyrate.
7 King's medium B slope (6.2.7).

Store the isolation plate at 2–4°C.

After 2 days incubation examine the re-streaked plates and test only sets of media that were inoculated from suspensions that gave only the original colony type on the re-streaked plate. Use the nutrient agar slope to perform oxidase (6.6.2) and tobacco hypersensitivity (5.2.1) tests. After 3 days read the O/F test and test for nitrate reduction. After 4 days observe the arginine dihydrolase reaction and test (6.1.5) for poly-β-hydroxybutyrate (PHB). After 3 and 7 days read the egg yolk reaction.

Presumptive diagnosis. a presumptive diagnosis of tomato pith necrosis can be made if bacteria with the following characteristics are isolated from tissue affected with typical symptoms of the disease:

1 Gram reaction negative.
2 Oxidative.
3 Green-fluorescent pigment not produced.
4 PBH accumulated (6.1.5).
5 Nitrate reduced *beyond* nitrite.

6 Lecithinase positive.
7 Arginine dihydrolase negative (weak positive reactions may be obtained if incubation is prolonged beyond 5 days.)
8 Tobacco hypersensitivity positive.

Confirmation of diagnosis. Confirmation of diagnosis can be made if injection of a suspension c. 10^8 cfu/ml (5.4) into the stem of half-grown tomato plants growing in humid chambers produces the pith necrosis symptoms. Host tests do not always readily give reproducible results but inoculation as described above usually produces pith necrosis and laddering within 2 weeks (5.3.4.1).

3.3.4 Xanthomonads

3.3.4.1 ***Xanthomonas campestris* pathovars**

The pathovars of *X. campestris* can only be differentiated by a host test.

Symptoms. Diseases caused by *X. campestris* exhibit a wide range of symptoms from leaf spotting and chlorosis (by far the commonest symptom) to cankers and even rotting and wilting. The use of the keys in Chapter 2 will often indicate the likely pathovar(s) involved.

Isolation. Make suitable suspensions from necrotic lesions or fresh or dried bacterial exudate (1.4), examine microscopically (1.3) and streak those containing bacteria on nutrient dextrose agar (6.2.10) or yeast dextrose chalk agar (6.2.20). Incubate for 3 days and look for smooth, yellow, domed, mucoid or viscid, rather slow-growing colonies (most are 1 mm diameter in 2–3 days). Colony colour varies usually from very pale yellow to a darker honey-yellow. Colonies of *X. campestris* pv. *manihotis* from cassava blight have no yellow pigment. A presumptive diagnosis can be made at this stage (see below).

Test procedure. Prepare a separate water suspension of each of 2–3 likely-looking colonies. From each suspension streak a nutrient agar plate and inoculate a set of the following media:
1 Nutrient agar slope (6.2.9).
2 Nutrient agar plates containing 0.1% and 0.02% triphenyl tetrazolium chloride (TTC) (6.6.5).
3 Potato plug (6.6.4).
4 O/F medium (6.3.1).

On the third day or when growth has appeared examine the re-streaked plates. If there are mixtures of colony types on any re-streaked plate(s) do not proceed with the tests on the set(s) of media inoculated from the suspension(s) used to streak them. For those sets that appear to be from pure cultures, use the slope culture to read oxidase (6.6.2) and catalase (6.6.1) reaction. Determine the presence of growth on the TTC plates. Determine the presence of growth and colony colour on the potato plug. Determine whether metabolism is oxidative or fermentative on O/F medium.

Presumptive diagnosis. With considerable experience, a presumptive diagnosis of a disease caused by a pathovar of *X. campestris* can be made if typical symptoms of the disease are present and if colonies characteristic of the species are dominant on the isolation plates. For those without such experience it is advisable at least to confirm that the typical colonies are those of a Gram-negative, short, rod-shaped oxidative bacterium.

If the method, starting to be used in diagnostic laboratories, for detecting the xanthomonadin pigments which appear to be unique to the genus is proved in practice, this would be a most useful and rapid adjunct to colony morphology for presumptive diagnosis. This method is given here because initial trial in our hands has shown that it is promising. Extract pigment in 3 ml methanol from nutrient agar cultures. Heat in a water bath at 90°C to facilitate extraction, centrifuge to remove cell debris, concentrate by evaporation at 90°C if necessary and scan using absorption spectrum analysis from 530–350 nm. A peak at 445 nm with shoulders at 467 and 420 nm (Fig. 3.1) indicates the presence of xanthomonadin pigments.

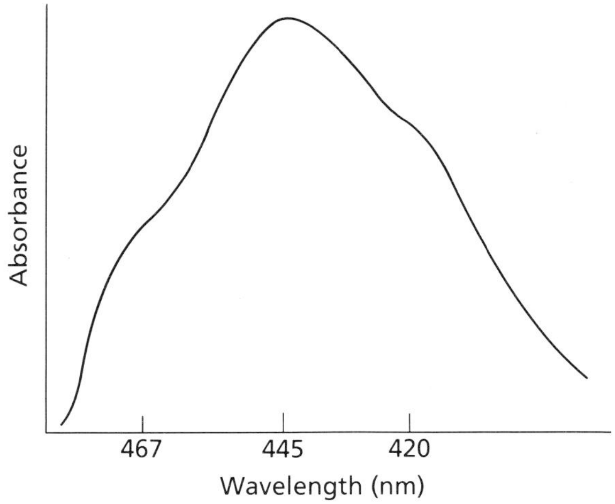

Fig. 3.1. Absorption spectrum for xanthomonadin pigment extract (*Xanthomonas campestri* pv. *campestris*).

Confirmation of diagnosis. Confirm a diagnosis of a disease caused by a pathovar of *X. campestris* if bacteria with the following characteristics are isolated from affected tissue with the general symptoms of this group of diseases:

1 Gram stain negative.
2 Oxidase reaction negative.
3 Catalase reaction positive.
4 Inhibition by 0.1% TTC.
5 Usually inhibition by 0.02% TTC positive/negative.
6 Slimy yellow growth on potato plug.
7 Oxidative metabolism.

Additional certainty of the diagnosis can be achieved if a host hypersensitivity test is positive (5.2.1). However, in the case of hypersensitivity tests on *X. campestris* isolates it should be noted that:

1 Not all pathovars give a positive reaction in tobacco, so use tomato as well (5.2.2).
2 The reaction is generally weaker than for fluorescent pseudomonads.
3 Tobacco plants should be held at 16°C for 4 days prior to inoculation and should be incubated at 33°C after inoculation.

A variety of host tests can be performed, including seedling tests, spray inoculations of young plants or fresh young organs (see Chapter 5, in particular 5.3.2 and 5.3.4). For example, young, detached walnut fruits may be used for *X. campestris* pv. *juglandis* (5.3.2.5) and *X. campestris* pv. *phaseoli* gives characteristic lesions on dwarf bean pods (5.3.2.6). Inhibition by TTC distinguishes xanthomonads from other yellow-pigmented bacteria commonly isolated from plant surfaces, e.g. *Flavobacterium* spp. and *Erwinia herbicola.*

Consideration of the host and symptoms involved in diseases from which *X. campestris* is isolated will allow diagnosis of the specific, pathovar-induced disease. For this refer to the lists published by Dye *et al.* (1980), to original papers and to the descriptions of selected major diseases given below. However, if specific quarantine and legislative questions demand confirmation, Koch's postulates need to be fulfilled. A wide selection of relevant host tests are given in Chapter 5.

*Black rot of crucifers (*Xanthomonas campestris *pv.* campestris*)*

Cabbage and cauliflower are commonly infected. Large, irregular, pale-brown lesions develop on leaves which often show sectored wilts, black veins and dwarfing. Infected leaves usually show marginal chlorosis and tattering. Vascular bundles in stems are discoloured and may exude a yellow, bacteria-laden ooze. In severe infections stem bases may rot. See also Hayward and Waterston, 1965d.

Bacterial blight, wilt, leaf spot of begonia (Xanthomonas campestris *pv.* begoniae)

Small, watersoaked spots develop, especially adjacent to the leaf margins. As they develop they become yellow or brown, infected veins usually collapse and in severe infections whole leaves may progressively wither and fall, sometimes causing death of the plant. See also Dye (1963), Taylor *et al.* (1981).

Citrus canker (Xanthomonas campestris *pv.* citri)

An extremely serious disease of most *Citrus* species. Having been eradicated from the USA it has recently been re-introduced. Several host-specific races occur.

On leaves, yellow spots on the undersurfaces turn brown and become rough and scab-like. Raised, corky lesions are found on stems. These spread rapidly in warm, moist weather. Scab-like lesions are also found on fruit. See also Hayward and Waterston (1964a), Stall and Seymour (1983).

Bacterial blight of hazel and filbert (Xanthomonas campestris *pv.* corylina)

In many orchards this is the most serious disease.

One of the most characteristic symptoms is necrosis of the emerging buds in late spring. Infected leaves show small, angular to irregular, brown or black spots. Black spots and streaks may be found on young stems and cankers may also be found on twigs and branches. Young, green nuts may also show small, black, necrotic spots. See also Locke and Barnes (1979), Wimalajeewa and Washington (1980).

Bacterial spot of pumpkin, squash and other cucurbits (Xanthomonas campestris *pv.* cucurbitae)

Often common in warm climates. Several cucurbits are susceptible including pumpkin, squash, melon, cucumber, marrow and courgette.

Small, yellow-brown, angular spots occur on leaves, coalescing to form irregular, necrotic spots. Watersoaked lesions on fruits develop rapidly in moist conditions, causing an extensive fruit rot, usually accompanied by exudations of yellow, bacteria-laden ooze. See also McLean (1958).

*Bacterial pustule of soybean (*Xanthomonas campestris *pv.* glycines*)*

Sometimes more important in tropical areas than bacterial blight caused by *Pseudomonas syringae* pv. *glycinea.*

Small, brown spots with yellow margins on leaves and occasionally stems, develop into raised, brown pustules. On leaves these may eventually collapse. See also Hedges (1924).

*Bacterial streak, wilt of grasses (*Xanthomonas campestris *pv.* graminis*)*

Ryegrass (*Lolium* spp.) appear particularly susceptible. There is some indication that most strains are host specific, although strains from ryegrass can produce symptoms in grasses from other genera by artificial inoculation. A common disease of ryegrass throughout many parts of Europe.

Plants may wilt, but initial symptoms are characteristic yellow streaks on the leaves. Inflorescences may become infected producing copious amounts of yellow bacteria-laden ooze in warm, wet weather. See also Egli *et al.* (1975), De Cleene *et al.* (1981), Van den Mooter *et al.* (1981) and Egli and Schmidt (1982).

*Bacterial leaf spot of ivy (*Xanthomonas campestris *pv.* hederae*)*

Initially small, green, watersoaked, angular spots develop on leaves, later becoming brown or black necrotic lesions sometimes with a chlorotic margin.

*Hyacinth yellows (*Xanthomonas campestris *pv.* hyacinthi*)*

A widespread disease in Europe but now rarely of economic importance. It is found largely on the florist's hyacinth (*Hyacinthus orientalis*) but has been isolated from other related species.

The most characteristic symptom is the yellow streaks produced in the bulb scales. These may lead to rot of the whole bulb. Infected bulbs may produce distorted leaves. Infected bulbs, when cut transversely, usually exude a yellow bacteria-laden ooze from which isolation should be made if present. Leaves may also develop long, narrow, watersoaked irregular stripes which are often more common towards the tips. See also Saaltink and Kamerman (1971), Beijer (1972) and Janse and Miller (1983).

*Bacterial blight of walnut (*Xanthomonas campestris *pv.* juglandis*)*

A common and widespread disease in most walnut-growing regions, sometimes serious.

The disease is characterized by black lesions, angular to irregular on leaves, long and narrow on young stems and irregular and often large on fruits. See also Miller *et al.* (1940), Bradbury (1967d) and Mulrean and Schroth (1982).

*Bacterial blight, angular leaf spot, black arm, boll rot of cotton (*Xanthomonas campestris *pv.* malvacearum*)*

Often a very serious disease of cotton. Many races are recognized on the basis of cultivar resistance. Leaf spots are small, angular and water-soaked, becoming dark-brown or reddish-purple as they age. Blighted leaves may absciss prematurely. Black, necrotic lesions may girdle and kill stems. Bolls show sunken dark-brown to purple or black lesions. Severely affected bolls may show stained lint. The disease can be particularly serious as a seedling blight. See also Hayward and Waterston (1964b), Ebbels (1976), Davis and Sandridge (1977), Bird *et al.* (1981).

*Black spot of mango (*Xanthomonas campestris *pv.* mangiferaeindicae*)*

Black, angular spots, often surrounded by chlorotic margins, are found on leaves. Raised, fissured, black, scab-like lesions occur on fruits and on stems black, raised cankers which produce yellow, bacteria-laden ooze in wet weather. See also Steyn *et al.* (1974), Daniel *et al.* (1975), Moffett *et al.* (1979).

*Cassava blight, leaf spot (*Xanthomonas campestris *pv.* manihotis*)*

A serious and widespread disease in most cassava-growing areas. The pathogen lacks the yellow xanthomonadin pigment characteristic of the species. *X. campestris* pv. *cassavae* also causes a leaf spot disease of cassava.

Symptoms caused by pv. *manihotis* are very variable. Watersoaked, angular spots occur on leaves. As these enlarge they coalesce, often causing premature abscission. Vascular infection may cause wilting, especially in young stems. See also Lozano (1975), Persley (1977), Maraite and Weyns (1979).

*Bacterial leaf blight, kresek disease of rice (*Xanthomonas campestris *pv.* oryzae*)*

This causes pale-green or greyish, watersoaked streaks predominantly at margins and tips which occur on leaves of young plants spreading throughout the whole leaf. The systemic infection (kresek), which is more prevalent in tropical regions, results in chlorosis, desiccation and death of young plants. For further details, see Bradbury (1970c), Lozano (1977), and Ou (1972).

*Bacterial leaf streak of rice (*Xanthomonas campestris *pv.* oryzicola*)*

Initially linear, narrow, interveinal, watersoaked, dark-green streaks occur on leaves. As these enlarge they coalesce and become yellowish-orange or brown. Small droplets of yellow ooze may be seen. The later symptoms are similar to those of bacterial leaf blight. For further details see Bradbury (1970d) and Fang *et al.* (1957).

*Bacterial blight, wilt, leaf spot of pelargonium (*Xanthomonas campestris *pv.* pelargonii*)*

This is a widespread and often serious nursery disease readily spread by use of the cutting knife during the taking of cuttings.

Symptoms are rather variable and may be found as a superficial or systemic infection. Small, often watersoaked, pale-brown spots appear on the undersurfaces of the leaves, later becoming darker brown and visible from the upper surface. Once bacteria gain access to the vascular tissues, wilting may occur, often seen as a sectoral wilt between the main leaf veins. Pale straw lesions may develop on stems. Vascular infections often become systemic and cut stems may show dark-brown to black internal necrosis, especially of the pith and cortex. Pockets of yellow ooze may also be found. See also Kivilaan and Scheffer (1958), McPherson *et al.* (1977).

*Common and fuscous blights of bean (*Xanthomonas campestris *pv.* phaseoli*)*

Often occurs in the same crop as halo blight from which it is sometimes difficult to distinguish, but tends to be more common in warmer climates. *Phaseolus vulgaris* is the major host but other beans (*Vigna* spp.) are also susceptible.

On leaves, initially small, watersoaked lesions develop narrow, yellow haloes. Spots turn brown as they enlarge and may coalesce. Severely infected leaves may appear tattered or may drop prematurely. Infected veins often appear watersoaked and may turn brick red. Reddish streaks may be found on the stems; as the plants grow these may split, exuding a yellow, bacteria-laden ooze in wet weather. In systemic infections the vascular bundles are usually discoloured and may produce ooze. On pods, watersoaked lesions surrounded by brick-red margins enlarge rapidly, often coalescing to form large, irregular, necrotic areas. Infected seeds are sometimes wrinkled and the hilum may be discoloured. See also Zaumeyer and Thomas (1957), Hayward and Waterston (1965e), Mazzucchi and Dalli (1975).

*Leaf spot, shot hole, black spot, bacterial spot of stonefruit (*Xanthomonas campestris *pv.* pruni*)*

Many *Prunus* spp. are susceptible but the disease is particularly important on plum, cherry and peach, especially in warm, temperate climates.

Small, chlorotic lesions occur on leaves. On the lower surfaces these have tan centres, becoming visible from the upper surfaces as they enlarge, then becoming darker, brown, maroon or black. Lesions are often surrounded by a chlorotic halo, may drop out to give a shot hole effect and tend to be more numerous at the leaf tips.

On peach stems, spring cankers which often enlarge to cause 'black tip', are found on the tips of over-wintering twigs. Summer cankers arise as watersoaked, purplish lesions around infected lenticels, later becoming dark and sunken. On plum and apricot stems, perennial cankers form; the inner bark is discoloured. On these hosts dieback is common.

Lesions on fruit of all hosts are similar — small, circular, dark-brown, sunken and pitted. On peach they are often surrounded by a pale-green halo. As fruits enlarge, the lesions crack and may exude a yellow, bacteria-laden ooze. See also Hayward and Waterston (1965f), Moffet (1973), Ercolani (1970).

*Bacterial leaf streak, black chaff of cereals (*Xanthomonas campestris *pvs.* undulosa *(on wheat),* hordei *(on barley),* secalis *(on rye))*

Originally grouped under *X. translucens*, these pathogens are now believed to be host specific and are given separate pathovar status. They are considered to be widespread but relatively unimportant pathogens. However, losses caused by them appear to be increasing, perhaps as a result of legislation against treating seed with organo mercury compounds; losses as high as 15% have been recently attributed to these pathovars on wheat and barley in East and Central Africa and the USA. Small, dark-green, watersoaked lesions develop on leaves. These develop into streaks often extending to the length of the leaf, becoming yellow-coloured and finally pale-brown. Ears may often be distorted and glumes often become watersoaked turning black (black chaff). See also Smith *et al.* (1919).

*Gumming disease (gummosis) of sugarcane (*Xanthomonas campestris *pv.* vasculorum*)*

This is a widespread and often serious disease.

Leaves develop yellow, narrow stripes with irregular margins. These stripes are more numerous at the leaf tips and as the necrotic areas enlarge, leaves wither and may die. Cut stems may show yellow to red-brown discoloration of the vascular bundles and the cut surfaces may exude yellow, bacteria-laden ooze. See also Hughes (1961), Bradbury (1973e).

*Bacterial spot or scab of tomato and pepper (*Xanthomonas campestris *pv.* vesicatoria*)*

Often an important disease of outdoor-grown tomatoes and peppers especially in the warm temperate climates.

Small, brown to black spots usually with chlorotic margins occur on leaves and stems. These may coalesce causing cankerous stem lesions, leaf blight and premature abscission. On fruits, spots appear as slightly-raised, corky scabs, usually irregular in shape, surrounded by a water-soaked margin. See also Hayward and Waterston (1964f), Mingtan and Watson (1973).

3.3.4.2 Leaf scald of sugar-cane (*Xanthomonas albilineans*)

This is primarily a vascular disease. The disease is widespread and often serious. A number of other *Gramineae* are known to be natural hosts.

Symptoms. Long, narrow, straight white stripes are found on leaves, often running much of the length of the leaf. In severe infections whole leaves may turn yellowish-white. Death usually starts at the leaf tips giving the plants a characteristic scalded effect. In severe infections the internodal distances are reduced and side shoot proliferation may occur. Cut stems usually show reddish-brown discoloration of the vascular bundles, especially around the nodes. See also Martin and Robinson (1961).

Isolation. Examine microscopically (1.3) and prepare suspensions from suitable plant material (1.4). Streak suspensions on sucrose peptone agar (6.2.18.1), incubate for up to 7 days and look for very small (<1 mm diameter) smooth, convex, entire, yellowish to yellow-buff, butyrous colonies appearing within 7 days.

Test procedure. From each of 2–3 likely-looking colonies prepare a separate suspension in sterile water. From each suspension streak 5% sucrose nutrient (6.2.17) and nutrient agar plates (6.2.9) and inoculate O/F medium (6.3.1). Incubate for 7 days. Examine the streaked plates. Carry out the following tests on culture from 5% sucrose nutrient agar plates that have only one colony type which is similar to that picked from the isolation plate (also read the O/F media inoculated from the same suspensions and record growth on nutrient agar); determine whether colonies are mucoid or butyrous; test oxidase reaction (6.6.2); remove and pool as much growth as possible from streaked plates of isolates that have: non-mucoid colonies on 5% sucrose nutrient agar, which do not grow on nutrient agar, and for which the corresponding O/F reaction is also oxidative. Test this growth for xanthomonadins (3.3.4.1).

Presumptive diagnosis. Make a presumptive diagnosis of leaf scald if bacteria are isolated with the following characters from affected tissue of plants with typical symptoms of leaf scald:
1 Mucoid growth on 5% sucrose nutrient agar.
2 No growth on nutrient agar.
3 Colonies as described above.
4 Xanthomonadins produced.
5 Oxidative.
6 Oxidase negative.
7 Gram reaction negative.

Confirmation of diagnosis. Confirm the diagnosis if bacteria are isolated with which the disease symptoms of leaf scald can be reproduced by host test (see 5.3.5.1).

3.3.4.3 Angular leaf spot of strawberry (*Xanthomonas fragariae*)

Commonly found in North America and recorded in Europe.

Symptoms. Small, angular, watersoaked, brownish spots appear on the undersurface of leaves, later becoming necrotic and visible on the upper surface. Runners and petioles may show reddish-brown spots. See also Kennedy and King (1962), Dye and Wilkie (1973), Bradbury (1977).

Isolation. Examine microscopically (1.3) and prepare suspensions from suitable plant material (1.4). Streak suspensions on nutrient dextrose agar plates. Incubate and after about 7 days examine plates for low-domed, smooth, entire, spreading, yellow-orange colonies. These tend to be less mucoid than *X. campestris*. A presumptive diagnosis can be made at this stage (see below).

Test procedure. Suspend each of 2–3 likely-looking colonies from the isolation plate separately in sterile water. Prepare Gram stains of each. From each that appears to be of solely Gram-negative, short rods, streak a nutrient dextrose plate (6.2.10) and a nutrient agar plate (6.2.9) and inoculate O/F medium (6.3.1). Incubate for 5–7 days and record type of growth on the plates and whether the reaction of O/F medium is oxidative or fermentative. Ignore the results from any suspension giving mixed colony forms on the nutrient dextrose plate. If facilities are available test for xanthomonadins (3.3.4.1).

Presumptive diagnosis. With considerable experience a presumptive diagnosis of angular leaf spot can be made if bacteria are isolated in large numbers from affected tissue of plants with typical symptoms of the disease that have the typical colony morphology of *X. fragariae*. Otherwise make a diagnosis if bacteria are isolated that also have the following characteristics:

1 Poor, pale-yellow growth on nutrient agar.
2 Oxidative.
3 Oxidase negative.
4 Gram reaction negative.
5 Xanthomonadins produced (if facilities are available).

Confirmation of diagnosis. Confirm the diagnosis if bacteria are isolated with which the disease symptoms can be reproduced by host test (5.3.5.1).

3.3.5 Coryneforms

Although most of these bacteria cause cankers and vascular diseases, leaf, stem and fruit spots are also often produced.

3.3.5.1 **Bacterial canker, bird's eye spot, or mealiness of tomato (*Clavibacter michiganense* subsp. *michiganense* syn. *Corynebacterium michiganense*)**

Symptoms. Raised, mealy lesions often occur on stems, leaves and fruits. Marginal leaf spots surrounded by a yellow halo and general leaf necrosis and scorching are common symptoms of the disease. The appearance of scorched windows in leaves is a result of a vascular infection; bacteria will not usually be isolated from such areas which occur most frequently in mature plants able to resist wilting.

Microscopic examination. Prepare suspensions from diseased tissue (1.4) and examine microscopically for the presence of bacteria (1.3). Prepare indirect fluorescent antibody stains (4.3.4) if the facility for this is available or Gram stains from suspensions in which bacteria are seen. Examine microscopically. For experienced diagnosticians further work may not be necessary and a presumptive diagnosis can be made at this stage.

Isolation. Streak suspensions containing bacteria (1.3) on nutrient dextrose agar (6.2.10) or yeast glucose mineral agar (6.2.21). Incubate for 3 days and examine plates for small, slow-growing, mucoid, yellow to pale-orange colonies.

Presumptive diagnosis. Depending on the experience of the diagnostician a presumptive diagnosis can be made if an IF test is positive, if affected tissue from plants with typical symptoms yield large numbers of Gram-positive coryneform rods or if colonies typical of *C. michiganense* subsp. *michiganense* and which contain only Gram-positive coryneforms are dominant on the isolation plates.

Confirmation of diagnosis. Confirm the diagnosis if a tomato cotyledon test (5.3.4.2) is positive. The presence of *C. michiganense* subsp. *michiganense* in the test cotyledons may be detected by means of the indirect fluorescent antibody test (4.3.4) or Gram stain (6.1.3). The disease is well-reviewed by Strider (1969). Other key references include Hayward and Waterston, (1964e) Volcani *et al.* (1970), Chaldecott and Preece (1983).

3.3.5.2 Leaf spot, wilt of bean (*Curtobacterium flaccumfaciens* pv. *flaccumfaciens* syn. *Corynebacterium flaccumfaciens*)

Symptoms. Although primarily a systemic, vascular disease, stems and pods can show rusty, brown lesions. Stem lesions are usually confined to the nodes. See also 3.5.8.

Microscopic examination. Excise leading edges of discoloured vascular tissue or necrotic lesions and macerate in sterile water. Examine microscopically (1.3) and Gram stain (6.1.3) suspensions in which bacteria can be detected. A presumptive diagnosis can be made at this stage.

Isolation. Streak suitable suspensions of macerated tissue (1.3 and 1.4) on nutrient dextrose agar (6.2.10). Incubate for 2–3 days and examine for slightly-raised, flat, often fluidal colonies. Almost all isolates are pale, creamy-yellow. A few may be orange or violet.

Presumptive diagnosis. A presumptive diagnosis of this disease can be made if Gram-positive, coryneform bacteria are detected in large numbers in affected tissue of a plant showing typical symptoms; or if such bacteria with typical colony morphology are isolated from such material.

Confirmation of diagnosis. The diagnosis can be confirmed if a host test with bacteria of the characters above is positive (5.3.5.1). For further information see Hedges (1926), Hayward and Waterston (1965a), Sinclair and Shurtleff (1975), Miller and Pollard (1976), Dinesen (1980).

3.3.5.3 **Silvering diseases of beet and tulip (*Curtobacterium flaccumfaciens* pv. *betae* and *C. flaccumfaciens* pv. *oortii* syn. *Corynebacterium betae* and *C. oortii*, respectively)**

Although these are primarily vascular diseases, leaf lesions are often produced from which the pathogen can be isolated.

Symptoms. There are two types of leaf lesions; silvery streaking along the leaf veins; silvery spots and roughened patches on the leaf.

Microscopic examination. Excise margins or underlying tissue from scabs or from soft, greyish tissue on bulbs (*C. flaccumfaciens* pv. *oortii*) or from the margins of silvery lesions on leaves (*C. flaccumfaciens* pv. *oortii* and *C. flaccumfaciens* pv. *betae*). Macerate in sterile water. Examine suitable suspensions microscopically (1.3 and 1.4) and prepare Gram stains (6.1.3) from suspensions containing bacteria. At this stage experienced diagnosticians can make a presumptive diagnosis (see below).

Isolation. Streak suspensions containing Gram-positive bacteria on nutrient dextrose plates (6.2.10) and incubate for 2–3 days. Examine for presence of cream to yellowish-cream smooth, flat, fluidal colonies with entire edges.

Test procedure. Prepare Gram stains from 2–3 typical colonies.

Presumptive diagnosis. A presumptive diagnosis of silvering disease can be made if Gram-positive coryneform cells are present in large numbers in suspensions from affected tissue from plants with typical symptoms of these diseases, or if such bacteria that have the typical colony form of these pathogens are isolated from such plants.

Confirmation of diagnosis. For *C. flaccumfaciens* pv. *oortii* inoculate leaves of tulip plants (5.3.4.5). Confirm the diagnosis if silvery, roughened patches containing masses of Gram-positive coryneform bacteria are produced. For *C. flaccumfaciens* pv. *betae* inoculate the leaves of red beet plants (Cheltenham Greentop) in the same manner (5.3.4.5). Again, confirm the diagnosis if silvery, roughened patches are produced containing masses of Gram-positive, coryneform bacteria. For further details of *C. flaccumfaciens* pv. *betae* see Keyworth *et al.*, 1956 and for *C. flaccumfaciens* pv. *oortii* see Saaltink and Maas Geesteranus (1969).

3.3.5.4 Bacterial blight of poinsettia (*Curtobacterium flaccumfaciens* pv. *poinsettiae* syn. *Corynebacterium poinsettiae*)

This bacterium causes a disease of poinsettia (*Euphorbia pulcherrima*) which is prevalent in the eastern states of the USA.

Symptoms. Stem and petiole spots and streaks with extensive watersoaking are the first symptoms. As the disease becomes systemic the cortex is discoloured yellow. Leaves and buds may become distorted. Necrotic marginal spots may also be found on the leaves.

Microscopic examination. Excise the margins of necrotic areas of cortical tissue or of necrotic lesions on stems and petioles. Macerate in sterile water and examine microscopically (1.3 and 1.4). Prepare Gram stains (6.1.3) from any suspensions in which bacteria can be readily seen. The white latex in host tissues may interfere with Gram staining. Experienced diagnosticians can make a presumptive diagnosis at this stage.

Isolation. Streak suitable bacteria-containing suspensions (1.4) on nutrient dextrose plates (6.2.10). Incubate for 3–5 days and examine plates for smooth, flat, slightly pink, fluidal colonies with entire edges.

Test procedure. Prepare Gram stains from 2–3 such typical colonies.

Presumptive diagnosis. A presumptive diagnosis of bacterial blight can be made if large numbers of Gram-positive, coryneform cells are detected in affected tissue of poinsettias with typical symptoms or if bacteria with such characters are isolated as the dominant organism on isolation plates from such tissues.

Confirmation of diagnosis. Prick stems and petioles of poinsettia plants. Paint over the pricks with a water suspension of test culture (10^5-10^6 colony-forming units/ml). Stab other plants to inoculate the vascular tissue (5.3.5.1). Incubate at *c*. 27°C in humid conditions. Symptoms may take more than a month to develop. Confirm the diagnosis if dark, necrotic lesions on stems and petioles or yellowing of vascular tissue around the point of inoculation are produced that contain masses of Gram-positive, coryneform bacteria. In the absence of host test results further confirmation may be obtained if typical bacteria tolerate 7% sodium chloride (6.6.6), grow at 36°C, hydrolyze aesculin (6.5.4) but do not produce urease (6.4.2). See also Starr and Pirone (1942).

3.4 **Cankers**

When certain pathogenic bacteria gain access to the woody tissue of the host, usually through wounds, their multiplication within the tissue may cause proliferation of plant cells and produce raised cankers. However, this term is often more loosely used — particularly for herbaceous hosts — for any surface lesion on stems that penetrates below the more superficial tissues. Proliferation may be a surface feature or a deep-seated effect in which the functions of the vascular system are impaired. Movement of bacteria in woody tissue is slow and cankers usually grow slowly. When deep-seated cankers girdle the stem, dieback usually occurs. Host tissue is not necessarily equally susceptible throughout the year and many cankers thus have an annual or seasonal cycle. It is not usually possible to isolate bacteria from cankerous tissue at all times or seasons. For certain cankers the margin between healthy and diseased tissue is sharply delineated, for others it may be diffuse.

Canker-forming bacteria belong primarily to three genera: *Pseudomonas* (green-fluorescent LOPAT group I), *Erwinia* (non-soft rot group) and *Xanthomonas*.

3.4.1 Preliminary schedule for the diagnosis of cankers

Microscopic examination. Cut into the lesion with a scalpel to determine the position of its margin and then make an incision in the epidermis of the healthy tissue parallel to and a millimetre or so from the margin. Lift the epidermis with the point of the scalpel and with blunt-nosed forceps, tear it back to expose the leading edge of the lesion. Dissect out part of the edge with a flamed scalpel and tease it out in sterile water

(1.3). Examine microscopically. The pathogen may only be present in small numbers, or in reasonable numbers only in small pockets of the tissue of dormant cankers. If bacteria cannot be detected immediately, it is often necessary to examine a number of the lesions microscopically to find these pockets, and even then their numbers may be too low to detect readily. It is often worthwhile sampling from a wide variety of samples and perhaps pooling them to facilitate isolation, although this is also likely to increase the number of non-pathogenic bacterial colonies on the isolation plates.

Isolation. Streak suspensions containing bacteria on the following media:
1 Nutrient agar plate (6.2.9).
2 Nutrient dextrose agar plate (6.2.10).
3 King's medium B plate (6.2.7).
4 Yeast dextrose chalk agar plate (6.2.20).
5 5% sucrose nutrient agar (6.2.17).
The growth on the isolation plates may enable a presumptive diagnosis to be made (see Table 3.3).

3.4.2 Bacterial canker of tomato (*Clavibacter michiganense* subsp. *michiganense* syn. *Corynebacterium michiganense*). See also 3.3.5.1

Within Europe the canker symptom is relatively uncommon but is often found in the USA.

Symptoms. Cankers arise from systemic infections in tomato stems. Pockets of bacteria develop in the vascular tissue and infect adjacent cortical tissue, which gives rise to an erumpent lesion in which the stems split. This is thus primarily a vascular infection and not a true canker.

Microscopic examination and isolation. Suspensions should be made from yellowed, vascular tissue, and the procedure is then as in 3.3.5.1. Isolation is often facilitated by placing cross-sectional discs of alcohol-flamed stems on nutrient dextrose agar (6.2.10). Characteristic yellow, fluidal colonies ooze from the upper cut surfaces of infected vascular bundles from which fresh plates of nutrient dextrose agar should be streaked.

Presumptive diagnosis and confirmation of diagnosis. As 3.3.5.1.

3.4.3 Fireblight (*Erwinia amylovora*) and Blossom Blight of Pear (*Pseudomonas syringae* pv. *syringae*)

The symptoms of these two diseases, particularly in the initial stages, are very similar and their host ranges overlap. Their diagnosis is therefore considered together.

Symptoms. In the early part of the season the only symptoms observed may be death of blossom with very little twig necrosis; therefore isolations from blossom stalks may be more productive than from twigs. Hold-over cankers which form in the late summer and often become dormant in the autumn may have diffuse edges. Bacterial multiplication may occur in autumn and even in winter, often, in the case of fireblight, with the production of ooze. In pear, fireblight usually causes a characteristic foxy-red discoloration of the wood underlying cankers. This discoloration may also be found in other hosts but is often less apparent. It does not occur with blossom blight. For more detailed accounts of symptoms of fireblight, Lelliott (1987), Van der Zwet and Keil (1979) and for blossom blight Garrett (1987) should be consulted.

Microscopic examination. Tease affected tissues in sterile water and examine microscopically (1.3) for the presence of bacteria after about 30 minutes. If available, inoculate immature pear fruits or slices of these with the teased tissue.

Isolation. Streak material in which large numbers of bacteria have been seen (1.4) on plates of 5% sucrose nutrient agar (6.2.17) and King's medium B (6.2.7). Incubate, examining the isolation plates daily for up to 5 days. Both *E. amylovora* and *P. syringae* pv. *syringae* produce levan-type colonies on 5% sucrose nutrient agar. Isolated colonies are usually 3–5 mm after 2 days and 5–7 mm after 3 days (Figs. 3.2 and 3.3) but may be slower growing — particularly from late or very early season samples. With experience, the two organisms can be distinguished by their colonial morphology on 5% sucrose nutrient agar. At 2 days both colony forms are radially striated, but typical *E. amylovora* colonies have steeper sides, a dense, flocculent centre and are often creamier in colour. Occasionally non-levan forms of *E. amylovora* are isolated — these should be expected if no levan colonies (6.3.9) appear, but ooze is produced on pear slice. Fast-growing *Bacillus* spp. occasionally swamp the 5% sucrose nutrient agar, and for this reason it is advisable to use the additional King's medium B isolation plate. This can also be used to differentiate the two pathogens as *P. syringae* pv. *syringae* — but not *E. amylovora* — produces a green-fluorescent diffusible pigment on this medium in 1–2 days.

Test procedures. If after 24 hours the growth on the 5% sucrose nutrient agar appears to be a pure culture showing incipient levan formation

Fig. 3.2. Levan production by *Erwinia amylovora* colonies on 5% sucrose nutrient agar for 2 days at 25°C (× 3 magnification). Crown copyright. Used by permission of the Controller of HMSO.

Fig. 3.3. Levan production by *Pseudomonas syringae* colonies on 5% sucrose nutrient agar for 2 days at 25°C (× 3 magnification). Crown copyright. Used by permission of the Controller of HMSO.

carry out a slide agglutination test on some of the growth with *E. amylovora* antiserum (4.3.5). A presumptive diagnosis can be made at this stage (see below).

If the test is negative examine the pear (if inoculated) for signs of milky ooze — this may not appear in its typical form until the third day of incubation. If such ooze is present on the fruit make a second attempt to isolate the organism by streaking a suspension of the ooze on 5% sucrose nutrient agar and King's medium B.

Occasionally strains of *E. amylovora* are isolated that do not readily produce ooze on pear fruits, but these are usually lysed by 'phage NCPPB 1509 and may or may not be lysed by 'phages NCPPB 1507 and 1508.

If agglutinable, levan colonies are not isolated, the pear test is negative and large numbers of fluorescent colonies are present on the King's medium B plate, it is possible that the agglutination-negative levan colonies are of *P. syringae* pv. *syringae*. Each of 2–3 such colonies should be suspended separately in sterile water and each suspension streaked on 5% sucrose nutrient agar and inoculated to a set of LOPAT media (3.3.2.1). Then proceed as in 3.3.2.1.

Presumptive diagnosis. A presumptive diagnosis of fireblight can be made if levan-type colonies that agglutinate with *E. amylovora* antiserum (4.3.5) are isolated from plants with symptoms of fireblight. This diagnosis has a high degree of certainty.

Confirmation of diagnosis. Confirm a diagnosis of fireblight if ooze is produced on pear slices (5.3.2.1) or symptoms produced on young pear shoots following inoculation (5.3.5.1).

A presumptive diagnosis of blossom blight can be made if bacteria with the LOPAT characters of group Ia are isolated in large numbers from plants with typical symptoms.

For further details of blossom blight see Crosse (1957), Garrett *et al.* (1966); Garrett (1987); for fireblight see Hayward and Waterston (1965b), Lelliott (1968), Van der Zwet and Keil (1979), Lelliott and Billing (1984).

3.4.4 Bacterial canker of stone fruits (*Pseudomonas syringae* pv. *morsprunorum* and *P. syringae* pv. *syringae*). See also 3.3.2.1

As well as causing cankers on the stems and branches of stone fruit trees (cherry, plum, peach, apricot and many other *Prunus* spp.) the bacterium also infects leaves, shoots and fruits (3.3.2.1). In the UK and in much of north-west Europe the morsprunorum pathovar is the usual cause of this canker, but in many other fruit-growing areas, the cause is the syringae pathovar.

The disease caused by the morsprunorum pathovar has a well-defined seasonal cycle with a winter canker phase alternating with a summer leaf disease. Cankers, which are not perennial, are formed in the autumn or winter, but do not increase much in size until the following spring, when they enlarge rapidly, killing large areas of green bark. Once blossom fall occurs the progress of cankers is arrested, the populations of bacteria within cankers decline and usually die out. At the same time the leaf infection phase occurs. Spur leaves tend to be resistant as soon as they mature, but young leaves on extension shoots are infected. However, bacteria multiply as epiphytes on all leaves during the summer and reside until leaf fall, being the main source of new canker infections via the exposed leaf scars.

The disease caused by the syringae pathovar has a somewhat different epidemiology. Cankers can be perennial and the bacteria overwinter in them as small populations. Rapid multiplication occurs in the spring, ooze is often produced and the bacteria spread to leaves by rain splash. Unlike *P. syringae* pv. *morsprunorum*, bacteria gain access to woody tissue only via wounds in leaf scars and bark, from which new cankers arise.

Symptoms. Symptoms vary between host species. In cherry, trees of all ages are susceptible and most cankers are found at sites of leaf scars on fruiting spurs. This usually results in dieback of the spur but may occasionally spread to form a canker in the parent branch. Cankers can also be located on the branches, especially in the crotch and angles between the branches. On younger, thin-barked branches cankers are first visible in the spring as shallow, discoloured, sunken bark lesions often showing the presence of amber-coloured, gummy exudates. When a canker girdles a branch, dieback will occur. In plum, cankers occur most frequently on stems and trunks, often leading to death of mature trees. Cankers may extend the length of the stem and often appear as dark-coloured, linear depressions in the bark. Once the stem is girdled, death ensues. Gumming is less common and not so obvious as on cherry.

Isolation. The bacterial populations in cankers are often too low to detect microscopically. Tease out affected tissues (1.3 and 3.4.1) and macerate in a little sterile water. Leave to stand for at least 30 minutes. Prepare Gram stains of each preparation and examine microscopically (1.3). Even if bacteria cannot be seen streak preparations from typically symptomatic tissues on plates of 5% sucrose nutrient agar (6.2.17) and King's medium B (6.2.7). Incubate 2–3 days and look for colonies which produce fluorescein on King's medium B and levan (6.3.9) on sucrose nutrient agar. Many strains of *P. syringae* pv. *morsprunorum* do not produce fluorescein on King's medium B although they may on other media (6.2.7 and 6.2.8).

Test procedures. As for *P. syringae* pv. *syringae* (3.3.2.1).

Presumptive diagnosis. A presumptive diagnosis of bacterial canker can be made when LOPAT Group Ia pseudomonads (fluorescent or non-fluorescent) are isolated in considerable numbers from affected tissue of plants with typical symptoms of the disease.

Because symptoms of these two diseases are similar, diagnosticians in countries in which both diseases occur and who are not fully familiar with their symptoms should carry out the tests shown in Table 3.4 to differentiate the two pathovars.

Confirmation of diagnosis. Confirm diagnosis if a host test (5.3.3.) with isolates with the above characters is positive at a stage when the host is susceptible.

Strains of *P. syringae* pv. *morsprunorum* are host specific. Isolates from cherry and plum will not cross-infect. Cherry isolates from the UK can be subdivided into two races on the basis of cultivar resistance. For further information on the disease see Crosse (1956 and 1957), Crosse and Garrett (1966), Garrett *et al.* (1966), Garrett (1987).

3.4.5 Citrus canker or citrus blast (*Xanthomonas campestris* pv. *citri*)

This disease is widespread and economically serious in many citrus growing areas but has been eradicated from others. All citrus species and hybrids are susceptible.

Symptoms. On leaves early symptoms appear as sunken, brownish-yellow spots on the under-surface. These later become visible from the upper surface and develop a rough, scab-like texture. On stems the disease appears as irregular, rapidly-spreading, raised, spongy cankers. The underlying tissue is a discoloured brown. On fruits small, corky lesions may appear.

Isolation. Excise small pieces of tissue from the margins of leaf spots and stem cankers and macerate in sterile water after washing in running water. Examine microscopically (1.3) and streak suspensions containing bacteria (1.4) on nutrient dextrose agar (6.2.10) or yeast dextrose chalk agar (6.2.20). Incubate and proceed as in 3.3.4.1.

Presumptive diagnosis and confirmatory diagnosis. As 3.3.4.1

3.4.6 Bacterial canker, spot and shothole of plum and peach (*Xanthomonas campestris* pv. *pruni*). See also under 3.3.4.1

This is an important disease of plums in certain areas but not of economic importance within western Europe. Almond, cherry and apricot are also susceptible.

Symptoms. Initial symptoms are typically angular leaf spots with chlorotic margins. Ooze may develop from them. Necrotic areas often drop out, leaving shotholes. On twigs, deep-seated, perennial cankers occur. Infected stems may become deformed and die. Dark-green, watersoaked blisters may develop on the tips of overwintering twigs. These enlarge and kill the growing shoots. Summer cankers appear as purplish, sunken lesions arising from infected lenticels.

Isolation. Wash cankers in running water. Excise leading edges (3.4.1), macerate in sterile water, examine microscopically and proceed as in 3.3.4.1.

Presumptive and confirmatory diagnosis. As 3.4.1. Perform a host test using healthy susceptible *Prunus* species (5.3.3 and 5.3.4).

3.4.7 Other cankers caused by *Xanthomonas campestris* pathovars

Hazel blight	*X. campestris* pv. *corylina.*
Walnut blight	*X. campestris* pv. *juglandis.*
Black spot of Mango	*X. campestris* pv. *mangiferaeindicae.*

Proceed as for citrus canker caused by *X. campestris* pv. *citri* (3.4.5 and 3.3.4.1). For host tests see 5.3.2.5 and 5.3.3.

3.4.8 Bacterial blight or black arm of grapevine (*Xanthomonas ampelina*)

Xanthomonas ampelina is an atypical xanthomonad and is perhaps not a valid member of the genus.

Symptoms. Grapevine blight is characterized by the development of small summer cankers on petioles, spurs, and young shoots. Spurs may initially be swollen and cracked and shoots have elongated chlorotic spots from which the cankers develop. Leaf spots may occur. Petiole infection may result in leaf distortion and necrosis. Infected buds may fail to open or may produce sickly, stunted shoots. Cankers may exude a whitish, bacteria-laden ooze. The disease is widely distributed and causes serious damage in most of the wine-growing areas of southern Europe and South Africa. It also occurs in Turkey and Afghanistan. See also Panagopoulos (1987), Bradbury (1973), Ridé and Marcelin (1983).

Isolation. Remove tissue aseptically from the leading edge of symptomatic lesions and macerate in sterile water. Examine microscopically (1.3) and streak suspensions containing large numbers of bacteria (1.4) on yeast dextrose chalk agar (6.2.20) or nutrient dextrose agar (6.2.10). Observe after 6–10 days for the presence of small, slow-growing, yellow to orange, convex, smooth, round, butyrous colonies. On nutrient dextrose agar, colonies appear punctiform. A presumptive diagnosis can be made at this stage (see below).

Test procedures. Gram stain (6.1.3) some typical colonies and suspend separately 2–3 of those that consist only of Gram-negative short rods in sterile water. From each suspension streak nutrient dextrose agar and nutrient agar plates and inoculate the following media:

1 O/F medium (6.3.1).
2 Urease medium (6.4.2).
3 Nutrient dextrose agar slope (6.2.10) (incubate in a waterbath at 36°C).

Incubate, and after 6 days perform an oxidase test from the nutrient dextrose agar plate (6.6.2). Observe growth at 36°C, record O/F test and urease reaction and record growth on nutrient dextrose and nutrient agar plates. Disregard results of any set of media inoculated from a suspension which gave rise to re-streaked plates that have mixed colony forms.

Presumptive diagnosis. A presumptive diagnosis of bacterial blight can be made by experienced diagnosticians if colonies with typical characters are dominant on isolation plates made from affected tissue of plants with typical symptoms. Otherwise, such a diagnosis can be made if bacteria are isolated from such tissue with the following characters:

1 Gram reaction negative.
2 No mucoid growth on nutrient dextrose agar.
3 Poor growth on nutrient agar.
4 Oxidative.
5 Oxidase negative.
6 No growth at 36°C.
7 Urease positive.

Confirmation of diagnosis. Confirm the diagnosis if typical symptoms are reproduced in a host test (5.3.4 and 5.3.3).

3.4.9 Poplar canker (*Xanthomonas populi*)

This disease was once associated with *Pseudomonas syringae*. The true causal organism was at first identified as an aplanobacter but is now classified as a cream-coloured, mucoid *Xanthomonas* sp. Cankers are hyperplastic.

The disease is common in many parts of Europe and many poplar species are seriously affected. It is most noticeable in spring and summer when large cankers form that produce masses of ooze containing the bacterium. This is spread by rainsplash, and the bacteria gains access to other trees via wounds and growth cracks. Wounded areas swell and buds above the infected site may fail to open in the spring. Some strains appear to be specific for willows (*Salix* spp.) A *Xanthomonas campestris* pathovar is reported to cause bark necrosis of poplar.

Symptoms. Cankers first appear as slightly swollen areas, often near leaf scars. These subsequently split longitudinally and develop rugose edges. Cankers develop slowly, often becoming very large and causing severe dieback and eventual death of infected trees. Cankers ooze copious quantities of a slimy exudate mainly in the spring but also in autumn, from which the bacterium is readily isolated on appropriate media. It may also be isolated at all times of the year from the cortical tissue underlying cankers. See also Ridé (1987).

Isolation. Macerate pieces of suitable cortical tissue or slime in sterile water (see above). Examine microscopically (1.3) and streak suitable suspensions (1.4) containing large numbers of bacteria on yeast peptone glucose agar (6.2.22). Incubate at 20°C. Examine plates after 3–4 days for smooth, cream, mucoid, opaque to semi-opaque growth. Often isolated colonies do not develop on primary isolation plates (see Fig. 1.3) and the bacterium dies quickly, usually in less than two weeks. If isolated colonies are not obtained suspend any such growth from areas of the wash-out streak apparently free of other colony forms in sterile water and re-streak on yeast peptone glucose agar and nutrient agar plates (6.2.9).

Test procedure. Examine a Gram stain (6.1.3) from the suspension(s). Test typical growth from the original plate(s) for oxidase (6.6.2) and catalase reaction (6.6.1). Incubate re-streaked plates at 20°C and examine at 3 and 7 days. Record growth. If the growth is still typical and apparently uncontaminated by other bacteria let the oxidase and catalase results stand. Otherwise repeat re-streaking or isolation until such growth is obtained; then do oxidase and catalase tests. Subculture from the yeast peptone glucose agar plates if a host test is to be made.

Presumptive diagnosis. A presumptive diagnosis can be made if bacteria with the following characters are isolated from affected tissue on plants with typical symptoms:
1 Gram reaction negative.
2 Oxidase reaction negative.
3 Catalase reaction positive.
4 No growth on nutrient agar.
5 Short, rod-shaped cells.

Confirmation of diagnosis. Confirm the diagnosis if a bacterium is isolated that will not grow on nutrient agar or in Hugh and Leifson's O/F + glucose medium (6.3.1), that gives a positive catalase reaction and a negative oxidase reaction, does not liquefy gelatin (6.5.2) or reduce nitrate (6.4.1). More certain confirmation is attained if a host test can be done (5.3.3). Confirm if swelling of the wounded tissue occurs within 2 months. If maintained under humid conditions, ooze should be produced from which *X. populi* may be re-isolated.

For further information on the disease see Bradbury (1970a), Burdekin (1972), Ridé and Ridé (1978), Ridé *et al.* (1978), De Kam (1980).

3.5 Vascular wilts and other systemic diseases

3.5.1 Introduction

Not all these diseases are truly wilts. Several diseases (which show the occasional presence of bacteria in the vascular system) are primarily diseases of the parenchyma and collenchyma tissue of stems.

Microscopic examination and isolation. The cortical tissue of the stem should be removed with needles, knives or scalpels and blunt-nosed forceps to expose the underlying vascular tissue. Look first at the base of the stem for tissue discoloration and then work upwards until it is less evident. Dissect out part of this less-discoloured vascular tissue. Suspend it in sterile water and tease it out to facilitate release of bacteria. Examine microscopically. If bacteria are not seen in reasonably large numbers work down towards the base of the stem until they are found before streaking on isolation plates.

Unless the symptoms strongly suggest a particular disease streak suspensions on plates of the following media:
1 Nutrient agar (6.2.9).
2 Nutrient dextrose agar (6.2.10) or yeast dextrose chalk agar (6.2.20).

3 5% sucrose nutrient agar (6.2.17).
4 King's medium B plate (6.2.7).
Incubate and refer to Table 3.3 for an indication of the possible pathogens involved.

3.5.2 Bacterial canker of tomato (*Clavibacter michiganense* subsp. *michiganense* syn. *Corynebacterium michiganense*). See also 3.3.5.1 and 3.4.2

Symptoms. This disease is now sometimes referred to as 'corynebacterial wilting disease' since wilting is a fairly common symptom of the disease; canker production is not. Leaf wilting is often unilateral, especially in young leaves. Young plants are more susceptible to wilting than older plants, affected young leaflets remaining small, wilted and then becoming necrotic. Where plants resist wilting, blockage of the vascular tissue often results in pale brown 'scorched' windows between the leaf veins. Developing, green fruits may appear marbled with longitudinal, chlorotic streaks. Systemic fruit infections often have masses of bacteria in the placenta giving the vascular tissue of the fruit an orange discoloration. Infected seeds are often shrivelled and black.

If the vascular phase of the disease is suspected proceed as follows: Remove wilted leaves and look for bacterial ooze at the bases of the broken petioles. If this cannot be found examine the vascular system as described at the beginning of the section looking for straw to creamy-yellow discoloration of the bundles. In advanced infections the presence of large quantities of bacterial ooze around the vascular bundles facilitates removal of the outer layers of stem tissue.

Microscopic examination. If discoloration of the original tissue or ooze is found, tease such tissue or suspend ooze in sterile water, examine microscopically (1.3) and streak any suspensions in which many bacteria were seen (1.4) on nutrient dextrose agar plates (6.2.10).

If discoloration or slimy ooze is not detected remove sample sections of stem from the entire length of the stem and petioles. Flame in alcohol, and slice aseptically. Place these slices on nutrient dextrose agar (6.2.10). Incubate and examine daily for 4 days for the presence of yellow, bacterial ooze from the cut surfaces of the vascular bundles. Re-streak any ooze on fresh nutrient dextrose agar to obtain isolated colonies, then proceed as in 3.3.5.1.

3.5.3 Silvering of beet (*Curtobacterium flaccumfaciens* pv. *betae* syn. *Corynebacterium betae*). See also 3.3.5.3

Dissect out, as aseptically as is practicable, part of the vascular systems of stems and roots, particularly any showing staining, discoloration or necrosis. Tease these in sterile water and proceed as in 3.3.5.3.

3.5.4 Silvering of tulips (*Curtobacterium flaccumfaciens* pv. *oortii* syn. *Corynebacterium oortii*). See also 3.3.5.3

Symptoms. When bulbs are cut transversely the vascular tissue may appear yellow. Often bulbs are inspected for this disease by cutting across the neck and looking for yellowing around vascular bundles; when this method is used it is necessary to distinguish yellow bacterial spotting from the developing anthers of the inflorescence. Hard, yellow scabs may also be present on the outermost fleshy scales. These may become brown and necrotic.

Microscopic examination. Tease suitable vascular tissue (1.4) or macerate yellow scabs in sterile water and examine microscopically (1.3). Prepare Gram stains (6.1.3) of any suspensions in which bacteria can be seen. Then proceed as in 3.3.5.3.

3.5.5 Bacterial blight of poinsettia (*Curtobacterium flaccumfaciens* pv. *poinsettiae* syn. *Corynebacterium poinsettiae*). See also 3.3.5.4

Symptoms. The systemic disease caused by this bacterium is characterized by yellowing and premature leaf abscission; the vascular and cortical tissues may be discoloured. Infected stems may twist inwards and downwards. These symptoms are usually accompanied by necrotic lesions, whitish blisters and marked watersoaking of the stems. Proceed as in 3.3.5.4.

3.5.6 Bacterial wilt of lucerne (alfalfa) (*Clavibacter michiganense* subsp. *insidiosum* syn. *Corynebacterium* insidiosum)

Symptoms. The pathogen typically causes stunting, dwarfing of the leaves and a witch's broom effect. In some varieties there is also a yellow-brown to orange discoloration of the vascular tissue of the outer wood of the collar and roots. In some varieties (e.g. Europe and Vertus) the organism is more easily isolated from the vascular tissue of stems (which is usually not discoloured) than from discoloured roots. For a full description of the disease and its symptoms see Lelliott (1987). Other key references include Renfro and Sprague (1959), Hayward and Waterston (1964e), Close and Mulcock (1972), Hale and Close (1974), Miller and Pollard (1976), Lelliott (1979).

Microscopic examination. Tease suitable vascular tissue (1.4) in sterile water and examine microscopically (1.3). Prepare Gram stains (6.1.3) of suspensions in which bacteria are detected. If facilities are available for IF microscopy also prepare IF stains (4.3.4) from the same suspensions. A presumptive diagnosis can be made at this stage.

Isolation. If further work is considered necessary streak plates of nutrient dextrose agar (6.2.10) or yeast dextrose chalk agar (6.2.20) from suitable suspensions (1.3) and incubate.

Examine plates after 3–5 days for the presence of colonies that are slow-growing, whitish at first becoming cream or pale-yellow, opaque and fluidal. Some soil organisms inhibit *C. michiganense* subsp. *insidiosum* which then appears as an almost translucent background growth.

The production of dark, bluish granules of indigoidine is characteristic of *C. michiganense* subsp. *insidiosum*, although English isolates are often slow to produce the pigment. To enhance indigoidine production, subculture isolates on nutrient dextrose and yeast dextrose chalk agar and incubate at 25–27°C until the culture is grown, then transfer to 15–20°C and to 4–5°C and incubate for up to 2 weeks until indigoidine appears.

The pigment granules, which are cubic and can be readily seen individually under a low-power stereo microscrope, are found mostly in the agar under the margins of confluent growth on plates. Economy in plates can be made if the primary isolation plates are incubated at normal temperature and, if indigoidine is not produced, suspensions of four colonies typical of *C. michiganense* subsp. *insidiosum* and containing only Gram-positive coryneforms (if present) are each streaked on one quarter of two nutrient agar and two yeast dextrose agar plates. After good growth has been made at 27°C one plate of each medium can be incubated at 4–5°C and the other at 15–20°C.

Presumptive diagnosis. A presumptive diagnosis of lucerne bacterial wilt can be made if Gram-positive, coryneform cells are demonstrated in large numbers from the vascular systems of plants with typical symptoms. So long as large numbers of such cells are seen in vascular tissue it is unlikely that they are saprophytic soil coryneform bacteria, which could occur, for example, where roots have rotted or have been wounded. A similar diagnosis can be made even where symptoms are atypical or absent, but it is preferable to have isolated bacteria with the typical colony characteristics of *C. michiganense* subsp. *insidiosum*. A more certain diagnosis can be made if an IF stain (4.3.4) from such material, whether typical, atypical or symptomless, demonstrates large numbers of IF-positive bacterial cells.

Simple slide agglutination (4.3.5) using dense, heated (100°C for 30 minutes) suspensions of typical colonies of Gram-positive coryneforms from isolation plates will also increase the certainty of diagnosis, but this method has the disadvantage that some cultures of *C. michiganense* subsp. *insidiosum* react weakly and need microscopic examination to detect agglutination.

Confirmation of diagnosis. Normally the presumptive diagnosis, particularly where positive serological reactions have been obtained, is sufficiently accurate not to need confirmation. However, if this is required inoculate young lucerne seedlings of a susceptible cultivar such as Du Puits or Grimm (5.3.5.4). Observe for wilting. For further details see McCulloch (1925). Confirm the diagnosis if the inoculated plants wilt within 2–5 weeks.

3.5.7 Bacterial ring rot of potatoes (*Clavibacter michiganense* subsp. *sepedonicum* syn. *Corynebacterium sepedonicum*)

This disease is prevalent in North America, Scandinavia and the USSR.

Although diagnosis of the disease in tubers with typical symptoms is reasonably straightforward, more advanced rots become contaminated with many saprophytic bacteria which can overgrow *C. michiganense* subsp. *sepedonicum* on isolation plates and which, in the case of lactobacilli, can be mistaken for the pathogen in a Gram stain. The bacterium also has a latent phase in which symptoms are absent, although bacterial populations may be high.

Symptoms. The disease in the field is usually seen as a progressive wilt of the plant. If wilted stems are cut across a few inches above soil level a milky exudate can often be expressed from the bundles by squeezing the stem near the cut surface with pliers. Tubers are infected via the vascular bundles in the stolons and only show external symptoms at advanced stages of infection. Even internal symptoms can either be so slight as to be unnoticed or, in the case of latent infection, absent. Although in severely affected crops, tubers will often show external and internal symptoms on lifting, these symptoms are more commonly found in tubers towards the end of the storage period because the disease continues to develop within the stored tuber.

To detect internal symptoms in tubers cut them transversely close to the heel end, and look for cheesy rot around the vascular bundles. If such rotted tubers are pressed with the thumbs against the skin of the tuber near the cut surface a pale, creamy-yellow, cheesey matter can be squeezed from the region of the infected bundles. In early stages of infection the only symptom may be a glassiness of the tissue around the vascular system which is best seen immediately below the point of stolon attachment. In advanced stages, areas of rot coalesce giving the characteristic ring rot symptom in which the cortex separates from the inner tissues.

Because of the difficulty of isolating and growing the pathogen, the most certain and sometimes quickest diagnosis is to inoculate young seedlings of eggplant with affected tissue or suspensions of such tissue taken from diseased plants. Such seedlings (see below) can form the basis of a diagnosis without isolation of the pathogen or can serve as a selective medium from which isolation into pure culture can subsequently be made. Laboratories that regularly handle ring rot infected specimens should have available supplies of seedlings at the correct stage throughout the months when such specimens are expected. For a fuller description of symptoms see Lelliott (1987).

Isolation and microscopic examination. Inoculate young eggplants (if available) with rotted material from the vascular rings of tubers or with the exudate expressed from the vascular bundles of wilted stems (see under 'confirmation of diagnosis' below). If no rotted material or exudate can be expressed macerate glassy or discoloured tissue from and around the vascular tissue in sterile water and use this to incubate eggplants. If suitable eggplants are not available, have some sown for use later in the procedure, store a selection of specimens at 2–4°C and macerate suitable affected tissue (see above) in sterile water. Make Gram stains (6.1.3) and if the facility is available, IF stains (4.3.4) from macerates. Examine the Gram and IF stains microscopically. A presumptive diagnosis can be made at this stage (see below).

Test procedure. If the diagnostic procedure needs to be continued, streak suspensions containing Gram-positive and/or IF-positive coryneform bacteria on yeast peptone glucose agar (6.2.22) or yeast glucose mineral agar (6.2.21). Incubate at 21°C and examine plates daily for approximately one month. Several plates of each medium should be inoculated in case of overgrowth of some by other bacteria. If no rapidly-growing colonies occur suspend part of the wash-out streak in water and re-streak on freshly prepared medium. *C. michiganense* subsp. *sepedonicum* colonies are small, very slow-growing, whitish to pale, creamy-yellow, domed, opaque and mucoid. As they age they may become slightly fluidal and coalesce. Eggplants provide a very useful selective medium, and macerates of symptomatic tissue from them usually yield large numbers of colonies on yeast peptone glucose agar or yeast glucose mineral agar.

Presumptive diagnosis. A presumptive diagnosis can be made if affected tissue from plants with typical symptoms yields numerous (more than about 200/microscope field at × 1000 magnification) Gram-positive, coryneform cells about 0.5–1.5 μm long. The diagnosis is more certain if this is supported by the demonstration of equally numerous IF-positive cells from such tissue. It can be still more certain if eggplants inoculated with affected tissue produce typical symptoms (see below) and numerous Gram-positive and/or IF-positive coryneform bacteria can be demonstrated in the vascular system of stems or petioles of affected eggplants (see below).

Gram staining can be variable. Cells from young tuber rots often appear coccoid becoming more coryneform-shaped, with typical pairs of elbowed cells showing bending division, after sub-culture. If ooze is smeared directly on to slides, cells do not necessarily stain or may appear weakly Gram-negative; Gram-positive cells usually appear weakly stained and granular.

Confirmation of diagnosis. In certain cases it may be desirable to confirm the diagnosis by isolation and identification of the pathogen.

Isolates with the colony morphology of *C. michiganense* subsp. *sepedonicum* should be purified by re-streaking and tested for production of typical symptoms in a host of the pathogen. Symptoms of ring rot in potato plants and tubers are difficult to obtain under most glasshouse conditions and can take some months to appear. The most reliable results may be given with a stem inoculation (5.3.5.1) of 15–20 cm long shoots on tubers.

A more rapid and reliable test (5.3.5.5) uses young eggplants (cv. Black Beauty) which are sensitive to very low populations of *C. michiganense* subsp. *sepedonicum*. Only young plants about 8–10 cm high approximately at the stage of having the third true leaf half expanded are sensitive (Lelliott & Sellar, 1976).

A typical symptom in eggplants is sectoral wilting, especially around the shoulders of the leaf. This may be preceded by large, greasy lesions in the interveinal areas (Fig. 3.4.) These, with a demonstration of the presence of large numbers of typical Gram-positive and/or IF-positive coryneform cells in the symptomatic tissues constitutes a positive result. It should be noted that inoculation with populations of more than 10^7 cells/ml of some other bacteria can cause xylem blockage and wilting symptoms not very dissimilar to those caused by *C. michiganense* subsp. *sepedonicum*, but without large numbers of Gram-positive coryneforms in the vascular system distant from the point of inoculation. Plants usually begin to wilt after 6–10 days, but occasionally may not show wilting for several weeks. Eggplants should be incubated for up to 40 days.

Examination of potato tuber lots for latent infections of ring rot. The presence of latent infection with ring rot in higher grade seed lines can cause serious problems to growers and certification authorities. A method for detecting its presence (but not its amount) in stocks is summarized below (for details see Lelliott & Sellar, 1976).

1 Wash tubers in lots containing 100 or 200 tubers and allow to air dry.

2 Remove the epidermis from around the heel ends and then aseptically remove a conical core of tissue containing the vascular bundles from the heel ends using a flamed knife (the tip of a domestic potato peeler designed to remove potato eyes is ideal). Cores should be approximately 1 cm in diameter and 0.7 cm deep.

3 Pool the cores and just cover with 0.05 mol/l phosphate-buffered saline, pH 7.2, containing 1% DC Silicone Antifoam MS A Compound (Hopkins & Williams), 0.05% Lubrol W Flakes (sterilized by autoclaving) and 0.1% tetra sodium pyrophosphate (sterilized by membrane filtration). Homogenize with a blender for 4 minutes in an ice bath. Decant the supernatant.

4 Grind any large pieces of tissue with carborundum powder with a sterile pestle and mortar, add to the supernatant and allow to settle for 20 minutes. Filter through a single layer of muslin. Spin the filtrate at no more than 200 ***g*** for 10 minutes to remove cell debris with minimal loss of coryneform cells.

5 Pour off the supernatant and spin it at 12 000 ***g*** for 15 minutes.

6 Reconstitute the pellet in a total volume of 0.5 ml.

7 Inoculate the suspended pellet into at least 25 eggplants at the third leaf stage (5.3.5.5) by brushing the re-suspended pellet material into slits made with a scalpel in the stem to expose the vascular tissue. Incubate at less than 30°C for 40 days. Observe daily after 7 days for grease spots and sectoral wilting.

An alternative method for detecting latent infection in potato lots, which has the advantage of being more rapid but which is less sensitive, is to proceed with steps 1–6 of the eggplant mentioned above and then make an IF test on the reconstituted pellet (4.3.4). Where sensitivity is not of prime importance, for example in surveys of ware stocks, this method can be used as a screen by which samples that give an IF reaction are check-tested in eggplant to eliminate the false positives that IF can give. For further details and information on the disease see Hayward and Waterston (1964d), Snieszko and Bonde (1943), Lelliott and Sellar (1976), Anon (EPPO) (1961), Mazzucchi *et al.* (1983), Lelliott (1987).

3.5.8 Bean wilt (*Curtobacterium flaccumfaciens* pv. *flaccumfaciens* syn. *C. flaccumfaciens*). See also 3.3.5.3

Symptoms. Initial symptoms of systemic infection are a reversible diurnal wilt of the leaves. In well-established infections leaves may become necrotic and absciss, or show chlorotic interveinal patches. Plants may shrivel and die. Pods may show watersoaked, erumpent lesions later becoming chlorotic. As pods ripen, lesions characteristically remain greenish. Infected seeds may be discoloured and produce a yellowish ooze. A colour mutant can produce violet seed discoloration. For leaf spot symptoms see 3.3.5.3.

Microscopic examination and isolation. Dissect out discoloured vascular tissue, macerate in sterile water and observe microscopically (1.3 and 1.4). Prepare Gram stains (6.1.3) and streak on nutrient dextrose agar (6.2.10), and King's medium B (6.2.7) from suspensions that contain large numbers of bacteria. A presumptive diagnosis can be made at this stage. Incubate the isolation plates for 3–4 days and look for the fairly slow-growing, flat, fluidal, yellowish colonies of *C. flaccumfaciens* pv. *flaccumfaciens.* Gram stain any such colonies. If no colonies are seen and if green-fluorescent colonies are apparent on the King's medium B plate suspect systemic infection with *Pseudomonas syringae* pv. *phaseolicola* (3.3.2.1).

Presumptive diagnosis. A presumptive diagnosis of bean wilt can be made if large numbers of Gram-positive, coryneform bacteria are observed in the vascular system of plants with typical symptoms of the disease. This diagnosis is more certain if typical colonies containing only Gram-positive coryneforms are also isolated.

Confirmation of diagnosis. Stab-inoculate a faintly turbid suspension of the bacteria (*c.* 10^7 colony-forming units/ml (see 5.4)) into the stems of young *Phaseolus vulgaris* plants. Confirm the diagnosis if wilting occurs in 10–14 days (5.3.5.1). See also 3.3.5.2.

3.5.9 Yellow slime diseases of cocksfoot (*Dactylis glomerata*) and other grasses (*Clavibacter rathayi* syn. *Corynebacterium rathayi*)

These diseases appear to have a wide distribution but are not of major economic importance in temperature regions. The pathogens involved with a similar wheat disease, *C. tritici* and *C. iranicum* (syn. *Corynebacterium tritici* and *Corynebacterium iranicum* respectively), appear to be closely related if not synonymous. The disease in wheat attributable to *C. tritici* is often referred to as ear cockle; see Bradbury (1973c), Gupta and Swarup (1968), Pathak and Swarup (1984). The following methods are for *C. rathayi* but are probably applicable to the other species.

Symptoms. Symptoms are characteristic: a yellow, fluidal, bacterial ooze exudes from infected inflorescences and often covers other parts of the plants, which become stunted and distorted. The vascular bundles usually show discoloration and pockets of ooze. Infection appears to be assisted by nematode vectors.

Microscopic examination. Suspend droplets of ooze in sterile water or remove small pieces of discoloured vascular tissue (1.4) and macerate in sterile water. Observe microscopically (1.3) and prepare Gram stains (6.1.3) from suspensions that contain many bacteria. A presumptive diagnosis can be made at this stage.

Isolation. If further investigation is warranted streak suspensions on Doepel's medium (6.2.4), yeast extract glucose mineral agar (6.2.21) potato dextrose agar (6.2.13) or nutrient dextrose agar (6.2.10). Look for slow-growing, small, yellow, domed, mucoid colonies with entire edges. Colour may vary from lemon-yellow to a dirty, dark orange-yellow. Growth can be extremely slow, pin-point colonies usually being visible after 4 days.

Test procedure. Prepare Gram stains (6.1.3) from colonies likely to be *C. rathayi.*

Presumptive diagnosis. A presumptive diagnosis of yellow slime disease can be made if Gram-positive coryneform bacterium are observed in large numbers in or on affected tissue of plants with typical symptoms of the disease. The diagnosis is more certain if typical colonies containing Gram-positive coryneforms are also isolated.

Confirmation of diagnosis. Obtain pure cultures. Suspend each of 2–3 colonies likely to be *C. rathayi* and that contain only Gram-positive coryneform cells separately in sterile water and from each suspension streak a nutrient dextrose agar plate and inoculate a set of the following media:

1 Nutrient dextrose agar slope (6.2.10).
2 Aesculin medium (6.5.4).
3 Christensen's urease medium (6.4.2).
4 Nutrient broth containing 5% NaCl (6.6.6).

Incubate the nutrient dextrose agar slope in a water bath at 37°C and the other media at 27°C. Examine after 6 days.

Confirm the diagnosis if bacteria if the following characteristics are isolated:

1 Gram reaction positive.
2 Coryneform cells.
3 No growth at 37°C.
4 No growth in 5% NaCl.
5 Urease negative.
6 Aesculin hydrolysed.
7 Typical colony characters (see above).

Further confirmation can be obtained from a host test by pricking inoculating bacterial slime from infected plants into leaf sheaths at the bases of young shoots of the host from which it was isolated. Infection following such inoculation can be erratic.

For further information on the disease see Bird and Stynes (1977), Bird (1981).

3.5.10 Watermark disease of Willow (*Erwinia salicis*)

This disease is common in England, Holland and probably much of western Europe. It has not been recorded elsewhere.

Symptoms. Symptoms include a reddening of leaves in early summer. These later turn brown but do not drop. During the red-leaf phase, watery exudates of bacteria can ooze from infected twigs. Wood from established infections will show the watermark symptom as a greyish discoloration of the previous year's growth, which may extend throughout the stem and roots. See also Preece *et al.* (1979).

Microscopic examination. Extract the sap from the cut ends of several disease shoots by applying pressure in a vice or with pliers to force the sap out; if twig exudates are present, suspend some ooze in sterile water. Prepare suspensions and examine them microscopically (1.3).

Isolation. Streak any suspensions containing large numbers of bacteria (1.4) on pectate gel (6.2.12) and potato dextrose agar (6.2.13), and on either nutrient dextrose agar (6.2.10) or yeast dextrose chalk agar (6.2.20).

After 3–4 days examine plates and look for one week for moderately slow-growing, yellowish colonies on potato dextrose agar and white colonies on nutrient dextrose agar or yeast dextrose chalk agar. Look for shallow craters on the pectate gel.

Test procedures. Prepare Gram stains from 2–3 typical colonies and suspend those that are wholly Gram-negative in sterile water. Inoculate the following media and re-streak on pectate gel:

1 O/F medium (6.3.1).
2 Nitrate medium (6.4.1).
3 Gelatin (6.5.2).
4 Potato plugs (6.6.4).

Incubate for 3 days and examine re-streaked plates. Disregard any sets of media inoculated from suspensions that give rise to mixtured cultures on the re-streaked plates.

Presumptive diagnosis. A presumptive diagnosis of watermark disease can be made if Gram-negative bacteria are isolated with typical colony morphology and the following characters from affected tissue of plants with typical symptoms:

1 Slow fermentative metabolism.
2 No reduction of nitrate to nitrite.
3 No gelatin liquefaction.
4 Yellow growth on potato plug.
5 Weakly pectolytic (shallow crater without liquefaction).

Erwinia herbicola may be isolated but can be distinguished from *E. salicis* because it is gelatin-positive and nitrate reduction-positive and non-pectolytic.

Confirmation of diagnosis. Confirm the diagnosis if isolated bacteria with the characters given above reproduce typical symptoms in a host test (5.3.4 and 5.3.5.1).

3.5.11 Bacterial wilt of cucumber (*Erwinia tracheiphila*)

Symptoms. The first symptoms are rapidly-spreading, dull-green patches on the leaves, followed by a typical leaf wilt. These eventually shrivel and dry up. Infected stems exude a milky ooze when cut. See also Bradbury (1970).

Isolation. Transfer discoloured tissue or ooze from cut stems directly on to nutrient yeast dextrose agar (6.2.11) and nutrient agar (6.2.9) and examine for up to a week. Look for colonies on nutrient yeast dextrose agar that are slow-growing, whitish, low-convex, smooth, round with undulate to entire edges. Suspend 2–3 such colonies separately in sterile water and prepare Gram stains (6.1.3) from them. *E. tracheiphila* grows little if at all on nutrient agar.

Test procedures. Streak nutrient yeast dextrose agar and nutrient agar plates and inoculate O/F medium (6.3.1) from each suspension that is wholly Gram-negative, and consists of straight rods singly or in pairs or sometimes short chains. Disregard any sets of media inoculated from suspensions that give rise to mixed cultures on the streaked plates. When adequate growth has been obtained perform oxidase (6.6.2) and catalase tests (6.6.1) from colonies on nutrient yeast dextrose agar. Record gas production and whether oxidative or fermentative on O/F medium. Record growth on nutrient agar.

Presumptive diagnosis. A presumptive diagnosis of bacterial wilt can be made if bacteria are isolated with the colony morphology given above that have the following characters from affected tissue from plants with typical symptoms:

1 Gram reaction negative.
2 Short rods, single or in pairs and sometimes chains.
3 Oxidase negative.
4 Catalase positive.
5 Fermentative.
6 Gas produced.
7 No or poor growth on nutrient agar.

Confirmation of diagnosis. Inoculate test plants by stabbing stems with culture growth or by spraying leaves wounded with a pin-point with a barely turbid suspension (10^7-10^8 colony-forming units (5.4)) of bacteria with the characters given above. Use *Cucumis melo*, *Cucumis sativa* and *Cucurbita maxima* (musk melon, cucumber and squash respectively). Incubate in humid conditions at 25°C. Confirm the diagnosis if wilting occurs in approximately 7–10 days. *E. tracheiphila* dies out rapidly on most growth media; it also loses virulence very rapidly when cultured on agar media.

3.5.12 Blackleg of potatoes (*Erwinia carotovora* and *E. chrysanthemi*)

These pathogens cause two distinct diseases of potato: tuber soft rot and the systemic vascular necrosis and wilt known as blackleg. The diagnostic procedures given below are only for blackleg.

In cool, temperate climates the field symptoms of this disease are caused almost exclusively by the specialized pathogen *E. carotovora* subsp. *atroseptica* and subsequent rotting of tubers in store is mainly caused by this pathovar; in the UK 80–90% of bacterial tuber rotting is attributed to it. In warm, temperate and sub-tropical climates *E. carotovora* subsp. *carotovora* and *E. chrysanthemi* can cause blackleg symptoms on the stems; subsp. *carotovora* is probably the commonest cause of tuber-rotting in warm temperate and, with *E. chrysanthemi*, in sub-tropical climates.

Virtually all potato tubers are contaminated by one or more of the soft rot erwinias — *E. carotovora* subsp. *atroseptica* in cooler areas, *E. carotovora* subsp. *carotovora* in intermediate climates. In Western Europe blackleg is said to be more common in heavy clay soils.

Symptoms. The first symptoms occur usually just before the haulm meets across the rows, and appear as stunted plants with chlorotic leaves. The upper leaves show typical epinasty in which they become stiff and erect with inrolling margins. The base of the stem just above and below soil level is usually discoloured black or dark-brown. The lesion becomes necrotic but remains firm. Not all potato stems on a plant may be infected. Later in the season the affected leaves turn brown and the haulm dies.

Other stem infections may be caused by all three erwinias. In warm climates, pre-emergence stem wilting, post-emergence stem rotting with no typical blackleg lesions and shoot rotting are often the most common systemic symptoms.

Microscopic examination. Strip the epidermis from the cortex of the stem at the junction of healthy and diseased tissue. Tease and suspend the exposed blackened cortex and any cortical tissue adhering to the epidermis in sterile water. Examine microscopically (1.3).

Isolation. Streak suspensions containing many bacteria (1.4) on pectate gel (6.2.12) or crystal violet pectate medium (6.2.12) and on well-dried potato dextrose agar (6.2.13). Incubate for 2–3 days, inspecting the pectate gel plates after 12 and 24 hours for the presence of liquefying colonies. Erwinias produce deep, pectolytic pits on pectate gel and the CVP medium; colonies of *E. carotovora* are irridescent, translucent and criss-crossed with internal markings in straight-sided pits on the latter medium.

Test procedures. Select 2–3 typical colonies from a pectate plate and prepare separate suspensions of each colony in sterile water. If the facilities are available, prepare IF stains (4.3.4) and a slide agglutination test (4.3.5) from each suspension, against antisera for *E. c.* subsp. *atroseptica*, *E. c* subsp. *carotovora* and *E. chrysanthemi.* A positive presumptive diagnosis can be made at this stage (see below) but if negative results are obtained in the serological tests *or* if the tests are not made, prepare a Gram stain (6.1.3) from each suspension. Inoculate a set of the following media and streak on King's medium B (6.2.7) plates from each suspension that comprises only Gram-negative, short rods singly or in pairs or sometimes in short chains:

1 Pectate gel (6.2.12), or potato slice (5.3.6).
2 O/F medium (6.3.1).
3 Nutrient agar slope (6.2.9).

If it is desired to differentiate these erwinia pathogens also inoculate:

1 Purple peptone broth containing lactose (6.3.4).
2 Purple peptone broth containing α-methyl glucoside (6.3.4).
3 Sucrose medium (6.3.8).
4 Nutrient agar slope (6.2.9) at 36°C in an accurate waterbath.

Record the results of pectolytic activity at 1–2 days and oxidase (6.6.2) and catalase (6.6.1) activity on nutrient agar at 2 days and oxidative or fermentative metabolism, growth at 36°C, acid from lactose and α-methyl glucoside, and reducing substances from sucrose all at 3 days. Examine the King's medium B plates and ignore results from any sets of media inoculated from the suspensions that give mixed colony forms on King's medium B.

Presumptive diagnosis. A presumptive diagnosis of blackleg can be made if a strong, positive slide agglutination is obtained or if an IF test demonstrates large numbers (more than about 200 cells/microscope field at × 1000) of positive, short, rod-shaped cells occurring singly or in pairs from affected tissue of plants with typical symptoms. Such results can, with experience, suggest which of the three possible pathogens are involved, but the distinction between them must always be to some extent tentative because antisera at present available are not wholly specific (for further detail see 4.3). At present some 40 different serovars are recognized; most subsp. *atroseptica* strains isolated from blackleg plants belong to serovar 1 (De Boer *et al.*, 1979).

Otherwise a presumptive diagnosis of blackleg can be made if bacteria with the typical colony morphology given above are isolated that consist of Gram-negative, short, straight rods occurring singly or in pairs and bacteria with the following characters are isolated from affected tissue from plants with typical symptoms:

1 Strongly pectolytic.

2 Fermentative metabolism.
3 Oxidase negative.
4 Catalase positive.

A presumptive attribution of the aetiology can be made if the subsidiary test results are as follows:

	E. c. subsp. *carotovora*	*E. c.* subsp. *atroseptica*	*E. chrysanthemi*
Lactose	+	+	−
α-methyl glucoside	−	+	−
Reducing substances from sucrose	−	+	−
Growth at 36°C	+	−	+

Occasional aberrant strains will almost certainly be found and the diagnostician should be prepared to look for overall similarities in these properties rather than expecting all strains to fall neatly into place.

It should be ntoed that some species of *Bacillus* can give a Gram-negative reaction under certain growth conditions, are pectolytic and one in particular, *B. polymyxa*, is not infrequently associated with rotting plant tissue, particularly in warm climates. The latter should not be confused with the soft-rotting Erwinias because it is larger (2–5 μm long), is a spore former and has a different colony form. Some Gram-negative pectolytic *Bacillus* spp. are small and do not readily produce spores; they are, however, not causes of rot. They may best be differentiated from the truly Gram-negative erwinias by a potassium hydroxide solubility test (6.1.3).

Confirmation of diagnosis. Confirm the diagnosis of blackleg if typical symptoms are produced in potato plants inoculated by stabbing isolates into the vascular tissue (beneath the 'wings') of the stem base (5.3.5.1). Host tests with isolates thought to be *E. carotovora* subsp. *atroseptica* should be made at a mean growing temperature of less than 15.5°C. For further information on blackleg see Perombelon and Kelman (1980) and Lelliott (1987).

3.5.13 Slow wilt or bacterial stunt of carnation (*Erwinia chrysanthemi* pv. *dianthicola*)

This disease was at one time common in western Europe but is now uncommon following the instigation of control measures at the propagation stage.

Symptoms. Infected plants appear unthrifty, lighter-coloured with a rather staring habit, flaccid and there is loss of bloom of the leaves. Plants become grey-green. Wilting does not always occur and plants may be stunted. A root rot accompanies the aerial symptoms but not basal stem rotting. The stems of infected plants are often extensively cracked, but this is an unreliable diagnostic character because cracking is not uncommon, though usually less severe, on healthy plants.

The inner wood of the infected stem is usually brown at the base. The discoloration tails off into streaks which become lighter, dry and frayed towards the top of the stem. Sometimes, and particularly in young plants, the wood is discoloured only in steaks at the stem base. Where the stem is cracked, the underlying wood is often disorganized, leaving a dry, necrotic pocket.

Microscopic examination. Break open the stem and tease out discoloured wood from the distal part of the stem in sterile water. Teased preparations (1.3) from affected plants show the bacteria mainly in the vessels, with little or no necrosis of surrounding tissue or dissolution of the vessels themselves. Heavily-infected groups of vessels can often be traced to the tops of the stems and laterals. Bacteria are densely packed in the vessels, devoid of Brownian movement, and embedded in a gel, which is often extruded from the broken ends of the vessels. Occasionally, and especially in young xylem, the bacteria lie with their long axes parallel to the sides of the vessels, and those in the extruded matter were similarly orientated. Teased preparations also show very light-brown rods of gel, about 50 μm wide and up to 500 μm long, containing massess of bacteria which appeared to be the contents of infected vessels from which the walls have been broken and pulled off. A presumptive diagnosis can be made at this stage (see below).

Isolation. If the bacterial masses, described above, can be seen under a microscope, place the slide on a damp filter paper in a Petri dish and leave at room temperature for at least 6 hours or overnight. Then streak suspensions on to well-dried plates of potato dextrose agar (6.2.13) and pectate gel (6.2.12).

Incubate and look after 2 days for liquefaction or craters in the pectate gel and after 3–5 days for the production of large, spreading, umbonate colonies with a pinkish centre and a lobed to coralloid or feathery margin — the so-called 'fried egg' colony — on potato dextrose agar (Lelliott, 1956).

Test procedures. Select 2–3 typical, well-separated colonies from the potato dextrose agar isolation plate and prepare a separate suspension of each in sterile water. From each suspension streak a potato dextrose agar plate and inoculate a set of the following media:

1 O/F medium (6.3.1).

2 Purple peptone both containing lactose (6.3.4).

3 Potato slice (5.3.6) or pectate gel (6.2.12).

4 Phenolphthalein diphosphate agar (6.6.3).

5 Tryptone medium for indole production (6.4.4).

Record the rotting of potato or liquefaction of pectate gel in 1–2 days; record phosphatase at 2 days and oxidative or fermentative metabolism, acid from lactose, and indole (or α-methyl indole) from tryptone at 3 days. Discard any set of results in which the potato dextrose agar plate spread from that same suspension indicates a mixed culture.

Presumptive diagnosis. A presumptive diagnosis can be made if the characteristic masses of bacteria in a gel-like matrix are found (see 'microscopic examination' above) in plants with typical symptoms. The diagnosis is more certain if bacteria with the typical colony morphology of the pathogen are the dominant organisms on isolation plates.

Confirmation of diagnosis. Confirm the diagnosis of slow wilt if bacteria are isolated that with typical colony morphology (see above) which rot potato slices completely (or nearly so) or produce deep pits on pectate gel in 2 days, are fermentative, produce phosphatase and indole or α-methyl indole and no acid from lactose in 3 days.

Fermentation of lactose by *E. chrysanthemi* is, typically, delayed, and rarely apparent within 3 days. Further confirmation can be obtained by standing rooted carnation cuttings, the root tips of which have had about 1 cm removed, in a turbid suspension (*c.* 10^8 colony-forming units/ml (5.4)) of the isolates for 5–10 minutes before replanting (5.3.5.3). Typical symptoms are produced in about 3 months if the isolate is *E. chrysanthemi* pv. *dianthicola.*

3.5.14 Bacterial blight of chrysanthemum and some other ornamentals (pathovars of *Erwinia chrysanthemi*)

A generalized blight of many glasshouse plants involving stem rot and collapse of plants can be caused by this bacterium. If conditions are unfavourable or in some other hosts, infection may not progress further than leaf or stem necrosis. Host specificity within *E. chrysanthemi* is discussed by Dickey (1979, 1981) and Samson and Nassan-Agha (1978).

Symptoms. The first symptom on chrysanthemum is usually a greyish lesion on the stem which becomes watersoaked and is associated with vascular browning and a pith rot. The disease can progress to affect all the tissues of the stem and lead to wilting and collapse of plants associated with extensive cortical rotting. Symptoms can be similar on *Dahlia* spp., *Dieffenbachia* spp., *Euphorbia pulcherrima*, *Philodendron* spp., *Syngonium* spp. and *Musa paradisiaca*.

Microscopic examination. Tease excised, discoloured vascular tissue in sterile water and examine microscopically (1.3). Discoloured tissue from the vascular system distal to the stem lesion is often devoid of bacteria.

Isolation. Streak any suitable suspensions containing many bacteria (1.4) on pectate gel (6.2.12) and well-dried potato dextrose agar plates (6.2.13). Look for liquefaction and craters in petate gel at 1–2 days and whitish, umbonate colonies with a pinkish centre and irregular margins after 4–5 days; the 'fried egg' colony on potato dextrose agar characteristic of pv. *dianthicola* (3.5.13) sometimes not produced by other pathovars.

Test procedures. As for slow wilt of carnations (3.5.13).

Presumptive diagnosis. A presumptive diagnosis of bacterial blight can be made if bacteria are isolated from plants with blight symptoms, that rot potato slices completely (or nearly so) in 2 days, are fermentative, produce phosphatase and indole and no acid from lactose in 3 days.

Confirmation of diagnosis. Confirm the diagnosis if symptoms of the disease are reproduced by isolates in plants of the host from which the isolates were made (5.3.5.1, 5.3.5.2 and 5.3.5.3).

3.5.15 Bacterial wilt of maize (*Erwinia chrysanthemi* pv. *zeae*)

This disease resembles in symptoms those caused by other pathovars of this species. It should be noted that pathovar differentiation in *E. chrysanthemi*, although often useful as a device for labelling disease agents, is probably a poor representation of its host specificity, in which there appears to be considerable overlap between cluster of strains (Dickey, 1981). For diagnosis see 3.5.13 and Hoppe and Kelman (1969).

3.5.16 Bacterial wilt of carnation (*Pseudomonas caryophylli*)

This disease was at one time a major and widespread disease in the USA but is apparently now far less common following its control at the propagation stage of production.

Symptoms. The symptoms are not dissimilar to those of slow wilt (3.5.13) but the wilt is more rapid and is usually accompanied in the later stages by a sticky, brown, basal, stem rot. In plants without basal rot, the cortical tissues at the base of the stem should be peeled away to reveal the vascular tissues. In affected plants these usually exhibit a brown necrosis, which is darker and more extensive than that associated with slow wilt and, unlike it, is sticky to the touch.

Microscopic examination and isolation. Tease small pieces of discoloured vascular tissue from stems in sterile water. Select such tissue from different parts of the stem including as high up the stem as discoloration is seen. Examine microscopically (1.3) and streak plates of potato dextrose agar (6.2.13) from suitable suspensions in which many bacteria are seen (1.4). Incubate and look for entire, raised, tan-coloured colonies on potato dextrose agar that become purple-brown with continued incubation. A presumptive diagnosis can be made at this stage (see below).

Test procedure. If present, suspend each of 2–3 well-separated colonies with the above characters separately in sterile water. Streak on a plate of King's medium B (6.2.7) and inoculate a set of the following media:
1 O/F medium (6.3.1).
2 Arginine medium (6.4.3).
3 Nitrate medium (6.4.1).
4 Ayers, Rupp & Johnson's medium (6.3.2), containing 0.5% β-hydroxybutyrate.
Disregard any set inoculated from a suspension from which the streaked plate has a mixture of colony forms.

After 3 days record the O/F reaction, arginine hydrolysis, nitrate reduction and the accumulation of poly β-hydroxybutyrate (6.1.5) and production of diffusible green-fluorescent pigment.

Presumptive diagnosis. A presumptive diagnosis of bacterial wilt can be made if bacteria that have the characteristic colony morphology are isolated from plants with typical symptoms. Diagnosis is more certain if bacteria that do not produce diffusible green-fluorescent pigments, are oxidative, produce arginine dihydrolase, reduce nitrate to N_2 or N_2O and accumulate poly-β-hydroxybutyrate are also isolated.

Confirmation of diagnosis. Confirm the diagnosis of bacterial wilt if typical symptoms are reproduced when carnations are inoculated by the method given in 3.5.12 (5.3.5.1). If inoculated plants can be maintained at about 30–33°C symptoms appear within 7–10 days and plants should be killed in about 2–3 weeks.

For further information on the disease see Dimock (1950), Hellmers (1958), Bradbury (1973).

3.5.17 Stewart's wilt of maize (*Erwinia stewartii*)

This is often a serious disease of maize, especially sweet corn varieties.

Symptoms. Two phases of the disease occur — a typical wilt and a leaf blight that may be compared with leaf blights caused by *Pseudomonas avenae* and *P. andropogonis*. The most characteristic and obvious symptoms occur on the leaves and appear as pale, yellowish-green, longitudinal streaks with wavy margins often extending the leaf length. Older lesions may dessicate and turn brown. In susceptible cultivars a vascular wilt of the whole plant may follow. Husks on the cobs may show small, watersoaked spots which may exude a bacterial ooze especially on the inner surface. Tassels may be produced early and appear bleached before premature withering and death. There may be droplets of ooze on flecked husks and leaf spots. Transverse stem and leaf sections usually exhibit vascular discoloration and production of a yellowish ooze.

Microscopic examination. Suspend droplets of ooze in sterile water or, if these are not present, cut tissue from the leading edge of fresh, yellowish-green leaf lesions or dissect out discoloured vascular tissue of leaves and macerate it in sterile water. Examine microscopically (1.3).

Isolation. Streak suspensions in which large numbers of bacteria can be seen (1.3 and 1.4) on nutrient dextrose agar plates (6.2.10). Incubate and look for slow-growing, small, yellow, flat to convex, transparent, butyrous or fluidal colonies with entire edges. Prepare Gram stains (6.1.3) from suspensions of 2–3 such colonies. A presumptive diagnosis can be made at this stage (see below).

If this is not possible or if more certain diagnosis is needed, and particularly when symptoms may be confused with other bacterial diseases of maize (*Pseudomonas andropogonis*, *P. avenae* and *Erwinia chrysanthemi* pv. *zeae*), use each suspension of colonies that appears from the Gram stain to comprise only Gram-negative, short, straight rods singly or in pairs or sometimes short chains to inoculate a set of the following media and streak a plate of nutrient agar (6.2.9):

1 O/F medium (6.3.1).
2 Nitrate broth (6.4.1).
3 Gelatin medium (6.5.2).

If any nutrient agar plate is a mixture of colony forms, disregard the set of media inoculated from the same suspension.

After 3 days read the O/F reaction, test for reduction of nitrate, and observe gelatin liquefaction. Perform Gram stain, oxidase (6.6.2) and catalase (6.6.1) tests from colonies on nutrient agar.

Presumptive diagnosis. A presumptive diagnosis of Stewart's wilt can be made if typical symptoms are present, especially where pockets of yellow ooze are detected in the vascular system, and if typical, yellow colonies are isolated which consist of Gram-negative, short, straight rods singly or in pairs.

More certain diagnosis can be made if the bacteria isolated have the following characters:

1 Gram reaction negative.
2 Short, straight rods.
3 Fermentative.
4 Oxidase negative.
5 Catalase positive.
6 Nitrate not reduced.
7 Gelatin not liquefied.

Confirmation of diagnosis. Confirm the diagnosis if the symptoms of Stewart's disease are reproduced by means of a host test on 8–14-day-old plants of a susceptible maize cultivar by either of the following methods. Inoculate hydathodes by incubating droplets containing approximately 10^7 bacteria/ml on the leaf tips of susceptible sweet corn cultivars at a high relative humidity; infected plants should produce leaf streak symptoms. Otherwise, perform a stem-stab needle puncture through a droplet of inoculum into the basal node, or introduce a droplet of inoculum into the stem with a syringe (5.3.5.1); infected plants should wilt within 14 days and show yellow pockets of ooze in the vascular tissue.

Further information on the disease can be obtained from Bradbury (1967a) Pepper (1967), Robert (1967), Gasperini *et al.* (1982).

3.5.18 Brown rot of potatoes, Granville wilt of tobacco, bacterial wilt, banana wilt or Moko disease (*Pseudomonas solanacearum*)

P. solanacearum is a heterogeneous species. Well over 200 hosts are known and specific strains show varying degrees of host specificity. The most commonly accepted subdivisions of the species are the races of Buddenhagen and Kelman (1964) based on pathogenicity and the biotypes of Hayward (1964) based on biochemical characters. The races of Buddenhagan and Kelman are:

1 Race 1 — infects most members of the *Solanaceae* and many other plants

2 Race 2 — infects triploid banana (causing Moko disease) and *Heliconia* spp.
3 Race 3 — infects potato and tomato (but may survive in other *Solanum* spp.)

The races may also be differentiated by the method of Lozano and Sequira (1970) using infiltrated tobacco leaves (see below). Race 1 causes a typical wilt, Race 2 induces a hypersensitive reaction and Race 3 causes a mild chlorosis in the infected area.

Races 1 and 2 are common in the tropics. The optimum temperature for growth varies between 33 and 37°C. Race 3 is found in the more temperate northerly and southern latitudes and also in temperate montane areas within the tropics. For most Race 3 isolates tested the optimum growth temperature is 27–28°C.

Hayward's biotypes (or using recommended terminology 'biovars') can be distinguished on the basis of utilization of three hexose alcohols and three sugars (see Table 3.5).

Races 1 and 2 both contain bacteria conforming to biovar I, III and IV. Race 3 corresponds to biovar II.

Symptoms. This disease causes a rapid and irreversible wilt under favourable conditions. If stems of wilted plants are dissected the vascular system can be seen to be discoloured and a milky ooze exudes from cut surfaces. Infected potato tubers usually show a cream to brown discoloration around infected vascular tissue when cut tranversely near the heel end; macerated tissue and bacterial ooze can be expressed from such vessels by squeezing the skin with the thumbs near the cut surface.

Microscopic examination and isolation. Remove ooze or discoloured vascular tissue and macerate in sterile water. Examine microscopically (1.3) and streak suspensions in which many bacteria are detected (1.4) on

Table 3.5. Biovars of *Pseudomonas solanacearum*

	Biovar			
	I	II	III	IV
Utilization of:				
Maltose	–	+	+	–
Lactose	–	+	+	–
Cellobiose	–	+	+	–
Mannitol	–	–	+	+
Sorbitol	–	–	+	+
Dulcitol	–	–	+	+

sucrose peptone agar (6.2.18.1) Incubate and examine plates after 2 days. On sucrose peptone agar most *P. solanacearum* cultures are pearly-white, flat and fluidal, often showing whorls which develop further on prolonged incubation.

Suspend each of 2–3 such colonies separately in sterile water, examine them for the presence of poly-β-hydroxybutyrate (6.1.5) and prepare a Gram stain (6.1.3). With experience a presumptive diagnosis can be made at this stage (see below). If no colonies with characteristic morphology are obtained on isolation plates, attempt further isolation from the original or from fresh plant material, but only if the original symptoms were typical of the disease.

Test procedure. If characteristic colonies contain Gram-negative, short, rod-shaped bacteria, but no poly-β-hydroxybutyrate granules are detected, subculture such colonies to 0.5% hydroxybutyrate medium (6.3.2) and grow in shake culture for 3 days and re-test for these granules. If further diagnostic work is considered necessary, also suspend each of 2–3 such colonies separately in sterile water and streak a nutrient agar plate (6.2.9) and inoculate O/F medium (6.3.1) from each suspension. Incubate and test oxidase reaction (6.6.2) after 2 days from the plate and read the O/F test after 3 days. Disregard any sets inoculated from suspensions of which the plates indicate a mixed culture.

If the diagnosis requires biovar determination purify the cultures on which the presumptive or confirmatory diagnosis was made and inoculate the biovar medium (6.3.5) supplemented with the following carbohydrates: maltose, lactose, cellobiose, mannitol, sorbitol, dulcitol. Also include a control containing no carbohydrates. Incubate at 30–32°C for 14 days and record results. A positive reaction is given by the production of acid detected by the production of a yellow colour. Determine the biovar from Table 3.5. If race determination is required a tobacco hypersensitivity test should be made (5.2.1).

Presumptive diagnosis. With experience, a presumptive diagnosis can be made if typical colonies (see above) are predominant on isolation plates from affected tissue of plants with typical symptoms; and if such colonies consist of Gram-negative, short, straight rods occurring singly or in pairs that contain poly-β-hydroxybutyrate granules. Otherwise, or if more certain diagnosis is needed, make a presumptive diagnosis if, in addition, isolates are oxidase-positive and are oxidative. Note that if poly-β-hydroxybutyrate cannot be detected in bacteria from typical colonies on the isolation plates they may still be detected in β-hydroxybutyrate medium (6.3.2).

Biovar and Race Determination. The biovar can be determined from the results of the carbohydrate reactions (see above and Table 3.5). The race can be determined from the hypersensitivity reaction as follows:

1 Chlorosis after 2 days followed by wilting after 7–8 days (Race 1).
2 Hypersensitive reaction after 12–24 hours (Race 2).
3 Chlorosis only after 2–8 days (Race 3).

Confirmation of diagnosis. The diagnosis can be confirmed if bacteria with the characters determined by the presumptive diagnosis are isolated that reproduce the typical symptoms of this disease after inoculation into host plants. Use tomato as a test plant if Races 1 and 3 are suspected and banana if Race 2 is suspected, and stab stems with suspensions (5.3.5.1). Tomato host tests should produce wilting within 2–8 days at 30°C. On banana, wilting normally occurs in 1–2 weeks at 30°C.

For further information on the disease see Kelman (1953), Buddenhagen (1961) and the extensive review by Buddenhagen and Kelman (1964).

3.5.19 Hyacinth yellows (*Xanthomonas campestris* pv. *hyacinthi*)

Symptoms. Bulb scales show a progressive yellow rot. Severely affected bulbs may produce distorted leaves. See also 3.3.4.1

Microscopic examination and isolation. Cut the bulb transversely near the nose and prepare water suspensions from any slimy, yellow vascular bundles in the exposed scale leaves. Examine microscopically and streak suspensions containing many bacteria on nutrient dextrose agar (6.2.10) or yeast dextrose chalk agar (6.2.20) plates. Look after 2–3 days for yellow, smooth, entire, domed, shining, opaque colonies. A presumptive diagnosis can be made at this stage.

Test procedure. See 3.3.4.1.

Presumptive diagnosis. See 3.3.4.1.

Confirmation of diagnosis. Confirm the diagnosis if bacteria are isolated that reproduce the disease on inoculation of hyacinth bulb through leaf scales (5.3.5.1).

3.5.20 Wilt, bacterial leaf spot, stem rot, bacterial leaf spot of zonal pelargonium, pelargonium blight (*Xanthomonas campestris* pv. *pelargonii*). See also 3.3.4.1

Symptoms. Leaves usually show a characteristic interveinal sectoral wilt.

Petioles and whole leaves may wilt. Cut stems may show dark-brown to black internal necrosis, especially of the pith and cortex, in which pockets of yellow bacteria-laden ooze may be found. Small, watersoaked, pale-brown lesions may develop on the undersurface of leaves, later becoming visible from the upper surface.

Isolation, test procedures and *diagnoses*. Make suitable preparations of discoloured vascular tissue and examine microscopically (1.3). Proceed as for 3.5.19 (see also 3.3.4.1). For host test see 5.3.4 and 5.3.5.1.

3.5.21 Gumming disease of sugar-cane (*Xanthomonas campestris* pv. *vasculorum*). See also 3.3.4.1

This bacterium causes a very widespread and serious bacterial disease of sugar-cane.

Symptoms. Early systemic symptoms include pale leaf stripes, turning yellow and sometimes red or brown. Masses of bacterial ooze may exude from the plant, sticking leaves and sheaths together and often preventing flowering. Cut stems show a reddish discoloration of vascular bundles which ooze a yellow bacterial exudate.

Isolation. Make suspensions of suitably teased discoloured vascular tissue and bacterial ooze. Examine microscopically (1.3) and streak suspensions in which bacteria were observed on nutrient dextrose agar (6.2.10) or yeast dextrose chalk agar (6.2.20). Incubate for up to 7 days. Examine isolation plates for rather slow-growing, small, pale-yellow, mucoid, entire, domed, opaque colonies on nutrient dextrose agar. Proceed as for 3.3.4.1 and 3.5.19. For host test see 5.3.5.1.

Diagnosis. As for 3.3.4.1 and 3.5.19.

3.6 Soft rots

A variety of bacteria are known to cause soft or wet rots and it is often impossible to suggest from the symptoms which pathogen is likely to be present in rotted tissue. Highly acid fruits are rarely affected. *Pseudomonas marginalis* and *Erwinia carotovora* subsp. *carotovora* are the most commonly-recorded causes of bacterial soft rots, but many other bacteria may be involved either singly or in combination. The following may not be a complete list.

1 *Pseudomonas* spp.

P. viridiflava (LOPAT group II)

P. marginalis and *P. fluorescens/putida* (LOPAT group IV)
P. aeruginosa
P. cepacia (non-fluorescent)
P. gladioli pv. *alliicola* (non-fluorescent)
2 *Erwinia* spp.
E. carotovora subsp. *carotovora*
E. carotovora subsp. *atroseptica*
E. carotovora subsp. *betavasculorum*
E. chrysanthemi (various pathovars)
E. rhapontici
E. herbicola
3 *Bacillus* spp.
B. polymyxa
B. macerans
B. subtilis
***Cytophaga* (*Flavobacterium*) spp.**
C. johnsoniae ('*F. pectinovorum*')
***Clostridium* spp.**

3.6.1 Schedule for diagnosis of generalized soft rots

Symptoms. A rapidly-spreading rot of succulent leaf tissues is common in wet, tropical climates and under glass but less so in other climates. Stems, mainly the parenchyma tissue, can be rotted. Storage tissues are usually extremely susceptible, and when rotten there is often discoloration of adjacent tissues.

Microscopic examination. Prepare suspensions for microscopic examination (1.3) and isolation (1.4) by cutting into the healthy tissue of storage organs (e.g. tubers, roots, corms, bulbs and fleshy leaves) with a knife nearly as far as the rotted area. Break open the specimen to expose the margin between healthy and rotted tissue. Rots of the stem pith can be treated similarly but cortical rots are often better handled as in 3.5.12. Make a microscopic examination of rotted tissue from the edge of, and also from a few millimetres inside, the lesion.

Isolation. Streak suspensions seen microscopically to contain bacteria on pectate gel (6.2.12), King's medium B (6.2.7) and nutrient agar (6.2.9). Examine the plates after 1, 2 and 3 days. If the dominant organism on King's medium B is a non-pigmented, non-fluorescent bacterium and the pectate gel shows the presence of deep, liquefied craters suspect a species of *Erwinia*. If pectate gel colonies produce shallow craters (sometimes difficult to see except with oblique light) and if the dominant organism on King's medium B is fluorescent suspect a fluorescent pseudomonad.

If the dominant organism on the pectate gel is yellow-pigmented and pectolytic suspect a *Cytophaga* (*Flavobacterium*) spp. *Clostridium* spp. will not grow on media incubated aerobically. *Bacillus* spp. usually produce large, mucoid, spreading colonies on nutrient agar which often adhere strongly to the surface of the medium.

Test procedure. As soon as well-separated craters are obtained on pectate gel, suspend each of 2–3 pectate-degrading colonies separately in sterile water, streak plates and inoculate or streak as appropriate the following media from each suspension:

1 Nutrient agar plate (6.2.9).
2 Pectate gel plate (6.2.12).
3 King's medium B plate (6.2.7) if a pseudomonad is suspected.
4 Starch plate (6.5.3).
5 O/F medium (6.3.1).

After 2 days prepare Gram stains and do an oxidase test (6.6.2) on well-separated colonies on the nutrient agar plates and record pectate liquefaction. After 3 days record presence of diffusible green-fluorescent pigment (if King's medium B was used), O/F reaction and test starch hydrolysis. Disregard any sets of media which have mixed colony forms on the streaked plates. If a *Bacillus* sp. is suspected, look and test for spores (3.7.3).

Presumptive diagnosis. Refer to Table 3.6. An appropriate presumptive diagnosis can be made if bacteria with the characteristics given are isolated from rotted tissue. However, it should be recognized that the soft rot may be secondary on some other cause such as frost injury, a primary fungal pathogen or wounding. If further diagnosis is needed refer to 3.6.2–5.

3.6.2 Soft rots induced by *Erwinia carotovora* and *E. chrysanthemi*

E. carotovora subsp. *carotovora* is the most common of the pathogens in this group and is with the soft-rotting pseudomonads, the usual cause of bacterial rotting of a very wide range of plants. (See also 3.5.12, 3.5.14 and 3.5.15) But in cool, temperate climates *E. carotovora* subsp. *atroseptica* more commonly rots potato tubers (see also 3.5.12). *E. chrysanthemi* is a more common cause of rot in tropical and subtropical climates and in glasshouse ornamentals, particularly of fleshy leaves and stems (see also 3.5.14 and 3.5.15). A soft-rotting pathogen of beet in California has been ascribed to *E. carotovora* subsp. *betavasculorum* (Thomson *et al.*, 1981.)

Table 3.6. Characteristics for presumptive diagnosis of soft-rotting bacteria

	Erwinia spp.	*Pseudomonas* spp.	*Bacillus polymyxa*	Pectolytic (*Flavobacterium*) *Cytophaga* spp.
Gram reaction	−	−	+*	−
Pigments	−	GF	−	Y
O/F reaction	F	0	F	0
Pectate degradation	++†	(+)‡	++	+§
Starch hydrolysis	−	−	+	−
Spore production	−	−	+	−

F = fermentative metabolism; O = oxidative metabolism; GF = green-fluorescent diffusible pigment produced; Y = non-diffusible yellow pigment produced.
*Usually positive; some strains are variable or sometimes negative but can be identified by their reaction to the potassium hydroxide solubility test (6.1.3).
†Deep pits usually with liquefaction.
‡Shallow craters without liquefaction.
§Shallow craters usually with slight liquefaction.

To differentiate the pathogens use the 'test procedures' and 'presumptive diagnosis' in 3.5.12, 3.5.14 and 3.5.15. These are summarized in part of Table 3.7. In using these methods it should be borne in mind that the individual isolates of each pathogen may be aberrant for one or two characters and that the diagnosis of an isolate should be based on its *overall* similarity to the characters given here.

3.6.3 Soft rots induced by *Bacillus* and *Clostridium* spp.

Both of these genera consist of spore-forming Gram-positive bacteria. *Clostridium* spp. have been implicated in soft rotting; they are anaerobic and cannot therefore be isolated under aerobic conditions, and their true role as primary pathogens is not yet understood. Their taxonomy is confused. Accurate diagnosis is not easy and is not covered here in detail. For further information see Lund (1972).

Several *Bacillus* spp. are pectolytic and have been implicated with soft rotting. These include *B. polymyxa*, *B. macerans*, *B. circulans*, *B. subtilis*, and *B. cereus*. But of these only *B. polymyxa* is a generally recognized cause of soft rot.

All but *B. subtilis* and *B. cereus* are often small, apparently Gram-negative rods that do not readily produce bacterial endospores. A pasteurization test (6.6.7) is advisable if spores cannot be seen using malachite green (6.1.6) or phase contrast microscopy. A potassium hydroxide solubility test (6.1.3) may be better than a staining procedure to determine

Table 3.7. Differentiation of fermentative soft-rotting bacteria

	Bacillus polymyxa	*E. carotovora* subsp. *carotovora*	*E. carotovora* subsp. *atroseptica*	*E. carotovora* subsp. *betavasculorum*	*E. chrysanthemi*
Spore production	+	−	−	−	−
Starch hydrolysis	+	−	−	−	−
Growth at 36°C	+	+	−	−	+
Reducing compounds from sucrose		−	+	−	±
Acid from lactose	+	+	+	+	−*
α-methyl glucoside		−	+	−	−
Indole production		−	−	−	+
Phosphatase production		−	−	−	+

*Gives a delayed positive but a few strains are positive in 3 days.

the Gram reaction. *Bacillus* spp. may, rarely, not produce acid from glucose and give a variable O/F reaction. They can be confused with Erwinia soft-rotting species from which they can usually be distinguished by starch hydrolysis; *Bacillus* spp. are usually able to degrade starch, soft rot erwinias can not. If spore formation is detected this is a certain difference from the erwinias. Use the methods in 3.7.3 to diagnose soft rot by *B. polymyxa*.

3.6.4 Soft rots induced by *Cytophaga johnsonae* (*Flavobacterium pectinovorum*)

Soil contains many flavobacteria and these are likely to be found in soft-rotted tissue. One, '*Flavobacterium pectinovorum*' was considered to belong to *Cytophaga johnsonae*, Christensen (1977), Lund (1969).

Confirmation of diagnosis can be made if yellow, oxidative soft-rot bacteria reproduce the disease symptoms in the host from which they were isolated (5.3.6 and Table 3.6).

3.6.5 Soft rots caused by green-fluorescent pseudomonads

Three LOPAT groups contain pectolytic soft-rot bacteria — groups II, IVa and IVb. Their differentiation is considered in detail in 3.3.2.2 and 3.3.2.3 which is only summarized here.

If suspected, select 2–3 isolated pectolytic colonies, prepare separate water suspensions of each and from each suspension streak a King's medium B (6.1.7) plate and inoculate the following media:

1 Nutrient agar slope (6.2.9).
2 5% sucrose nutrient agar (6.2.17).
3 Potato slice (5.3.6) or pectate gel (6.2.12).
4 Arginine medium (6.4.3).

After 2 days test the nutrient agar slope culture for oxidase activity (6.6.2) and use the rest of the growth for a tobacco hypersensitivity test (5.2.1). Record the pectolytic activity. After 3 days record growth on sucrose nutrient agar (6.3.9) and the arginine dihydrolase reaction.

Presumptive diagnosis. Diagnose the appropriate pseudomonad if bacteria with the characters given in Table 3.8 are the dominant bacteria on isolation plates from soft-rotted tissue.

Confirmation of diagnosis. Confirm the diagnosis if the symptoms of the disease are reproduced in the original host (5.3.4) with isolates with the other characters described here.

Table 3.8. Differentiation of soft-rotting fluorescent *Pseudomonas* spp.

	Levan	Oxidase	Pectolytic activity	Arginine dihydrolase activity	Tobacco hyper-sensitivity
Group II (*Pseudomonas viridiflava*)*	−	−	+	−	+
Group IVa (*Pseudomonas marginalis*)	+	+	+	+	−
Group IVb	−	+	+	+	−

**Pseudomonas viridiflava* colonies have green centres when cultured on sucrose nutrient agar, although this character may be lost on subculture.

3.6.6 Diagnosis of other rot diseases

3.6.6.1 Crown rot of rhubarb and pink grain of wheat and internal browning of hyacinth (*Erwinia rhapontici*)

Although first associated with a disease of rhubarb (*Rheum rhababarum*) there is now evidence that this bacterium is a less specialized pathogen and probably part of the epiphytic flora of plants. It is known to cause pink grain of wheat and a rot of hyacinth (Roberts, 1974; Sellwood and Lelliott, 1978).

Symptoms. Crown rot of rhubarb is characterized by a brownish discoloration of the rhizome and often the bud tissue, which remain fairly firm. Affected wheat grains show a light-pink discoloration and a slight softening of the endosperm. Affected hyacinth bulbs have a dry, brown, internal necrosis extending into the bulb from the neck.

Isolation and test procedures. Prepare suitable tissue macerates in sterile water and examine microscopically (1.3). Those which show the presence of bacteria under microscopic examination (1.4) should be streaked on plates of:

1 Pectate gel (6.1.12).
2 Sucrose peptone agar (6.2.18.2).

After three days look for pale-pink, fluidal colonies on sucrose peptone agar and for cratered colonies on pectate gel. A presumptive diagnosis can be made at this stage (see below). If further investigation is necessary, suspend each of 2–3 likely-looking colonies from the sucrose peptone agar plate separately in sterile water and from each suspension streak nutrient agar and pectate gel plates and inoculate:

1 Rhubarb rhizome (if available) (5.3.6).
2 Onion slice (5.3.6).
3 Potato slice (5.3.6).
4 Cucumber slice (5.3.6).
5 O/F medium (6.3.1).

Incubate and after 2 days test the oxidase reaction (6.6.2) of colonies on the nutrient agar plate. Test for rot on the potato and cucumber slices and look for pink discoloration of the onion slice and for black discoloration and pith rot of the rhubarb rhizome. Disregard sets if the suspensions used to inoculate them gave mixed colony forms when streaked on nutrient agar. After 3 days record the reaction on O/F medium and liquefaction of pectate gel.

Presumptive diagnosis. A presumptive diagnosis of crown rot of rhubarb or pink grain of wheat can be made if bacteria that have typical colony characters of *E. rhapontici* on sucrose peptone agar and do not produce craters on pectate gel are isolated from plants with typical symptoms.

Confirmation of diagnosis. Confirm the diagnosis of the appropriate disease if the bacteria isolated from plants with characteristic symptoms have typical colony morphology, do not degrade pectate gel but cause a slow rot of potato and cucumber slices, produce a pink discoloration of onion slices and black discoloration and a dry necrosis of rhubarb rhizome (if done), are oxidase negative and fermentative.

3.6.6.2 **Bacterial rots of onion**

Although typical soft rot erwinias (e.g. *E. carotovora* subsp. *carotovora*) may occasionally cause onion rots, several other bacteria rot onions and other bulbs. These include two non-fluorescent pseudomonad species — *Pseudomonas cepacia* and *Pseudomonas gladioli*. *P. gladioli* is sub-divided into two pathovars: pv. *gladioli* and pv. *alliicola*, the latter being an important pathogen of onions in certain areas.
Symptoms. P. cepacia causes sour skin of onions, being characterized by a slimy, yellow rot of the outer, fleshy scale leaves. *P. gladioli* pv. *alliicola* causes a softening of the outer scales often with dark discoloration. *E. herbicola* is often associated with storage rots of onions, especially when curing temperatures have been high. For further details of sour skin disease see Burkholder (1950) and Bazzi (1979); for bulb rot by *P. gladioli* pv. *alliicola*, see Tesoriero *et al.* (1982) and for foliar diseases of onion by *Erwinia herbicola*, see Hattingh and Walters (1981).

Isolation. Excise small pieces of onion scale tissue from the leading edge of the rot, macerate in sterile water and streak those preparations seen by microscopic examination to contain bacteria (1.3 and 1.4) on plates of the following media:

1 Pectate gel (6.2.12).
2 King's medium B (6.2.7).
3 Nutrient dextrose agar (6.2.10).
4 5% sucrose nutrient agar (6.2.17).

Incubate for 3 days and record colony characters.

Test procedure. If the dominant colonies on plates are not fluorescent on King's medium B, degrade pectate gel and are flat, wrinkled yellow/brown to dark-brown with green-yellow to brown diffusible pigments on the three agar media (characteristics of *P. cepacia* and *P. gladioli*) test such colonies from the 5% sucrose nutrient agar plate for poly-β-hydroxybutyrate granules (6.1.5); if none can be detected, inoculate β-hydroxybutyrate medium (6.3.2) from the colonies, incubate and again test for the polymer. If the dominant colonies are yellow or white and are large, mucoid, domed, usually with denser whorls inside the colony, but do not produce diffusible pigments or degrade pectate gel (characteristics of *E. herbicola*), suspend each of 2–3 such colonies separately in sterile, distilled water and from each, streak a nutrient agar plate and O/F media (6.3.1). Incubate and after 2 days test colonies on the nutrient agar for oxidase reaction (6.6.2) and after 3 days record reaction of O/F medium. Disregard the result from any set if the plate streaked from the same suspension has a mixture of colony forms.

Presumptive diagnosis. A presumptive diagnosis of soft rot of onion caused by a non-fluorescent pseudomonad can be made if bacteria are isolated that produce shallow craters in pectate gel and produce non-fluorescent, greenish-yellow to brown diffusible pigment on nutrient dextrose agar and King's medium B, are oxidative and oxidase positive and accumulate poly-β-hydroxybutyrate.

On sucrose nutrient agar, *P. gladioli* usually produces a brown, diffusible pigment which is often delayed, whereas *P. cepacia* produces a yellow pigment; on nutrient dextrose agar *P. gladioli* colonies are usually wrinkled and may produce a diffusible yellow-green pigment. These differences are not sufficiently reliable to differentiate the two pathogens. Such differentiation can be made if typical symptoms of either disease were present in the original specimen.

A presumptive diagnosis of soft rot associated with *Erwinia herbicola* can be made if bacteria that have the following characters are isolated:

1 Typical colony characters (see under 'Test procedure' above).

2 No pectolytic activity.
3 Oxidase.
4 Fermentative.
5 Yellow pigment produced. Non-pigmented forms are sometimes isolated.

Confirmation of diagnosis. Confirm the diagnosis if the isolates of the bacterium presumptively diagnosed produce the symptoms seen in the original specimen in a host test on onion (5.3.6).

3.7 Mushroom diseases

3.7.1 Brown blotch (*Pseudomonas tolaasii*) and ginger blotch (*P. gingeri*)

Brown blotch is a widespread and serious disease of cultivated mushrooms; ginger blotch has been recorded only occasionally and all known records are from the UK. For further details of these diseases see Wong and Preece (1979), Wong *et al.* (1982), Fahy *et al.* (1981). *P. tolaasii* and *P. gingeri* are closely related.

Symptoms. Brown and ginger discoloured and necrotic blotches develop on the mushroom caps. Brown blotch symptoms develop more rapidly and infection penetrates further into the cap.

Isolation. Place a drop of sterile water on to the discoloured area of the cap. Streak immediately from it on plates of King's medium B (6.2.7) and nutrient agar (6.2.9). Incubate and look for fluorescent colonies on King's medium B and small, whitish, round, entire colonies on nutrient agar. A presumptive diagnosis can be made at this stage (see below).

Test procedures. If confirmation of the diagnosis is required, make separate suspensions of each of 2–3 of the dominant green-fluorescent colonies from the King's medium B plate in sterile water, re-streak from each suspension on 5% sucrose nutrient agar (6.2.17), nutrient agar (6.2.9) and pectate plates (6.2.12), and from each also inoculate a set of the following media:
1 Egg yolk medium (6.5.1).
2 Arginine medium (6.4.3).
3 Gluconate medium (6.3.6).

Incubate, and after 2–3 days examine the plates. Disregard results from any set(s) if the plates streaked from the same suspension is a mixture of colony types. For other sets record at 2 days levan production on 5% sucrose nutrient agar and pectolytic activity on pectate gel, test for oxidase reaction (6.6.2) and make a suspension for a hypersensitivity test (5.2.1) from growth on the nutrient agar plates. At 3 days record arginine dihydrolase activity and test for 2-keto gluconate and lecithinase activity. Perform a tobacco hypersensitivity test (5.2.1).

Presumptive diagnosis. A presumptive diagnosis of brown blotch can be made if green-fluorescent, non-mucoid colonies are dominant on isolation plates from affected tissue of caps with typical symptoms of the disease. Similarly, make a presumptive diagnosis of ginger blotch if green-fluorescent mucoid colonies are dominant and symptoms typical. If more certain diagnosis is required diagnose brown or ginger blotch as appropriate if bacteria with the following additional characters are isolated:

	Brown blotch	Ginger blotch
Levan-type colonies on 5% sucrose nutrient agar	−	−
Mucoid colonies on King's medium B	−	+
Oxidase	+	+
Pectinase	−	−
Arginine dihydrolase	+	+
Hypersensitivity in tobacco	−	−
2-ketogluconate	+	−
Lipase	+	−

These two bacteria cannot be differentiated from the commonly-occurring saprophytes of the *P. fluorescens*/*P. putida* complex except by a host test (5.3.8), although it would not be expected that these saprophytes would be the dominant organism in lesions similar to those of these diseases but caused by some non-pathogenic disorder. *Pseudomonas reactans* (inv.) a member of the same group may, however, be found in considerable numbers on mushroom caps. It is non-pathogenic and may often react with *P. tolaasii* by the formation of a white precipitate in some agar media between adjacent colonies. For further details see Wong and Preece (1979) and Cutri *et al.* (1984).

Confirmation of the diagnosis. Confirm the appropriate presumptive diagnosis if bacteria with the characters given above are isolated that reproduce typical symptoms of these diseases in a host test (5.3.8).

3.7.2 Drippy gill (*Pseudomonas agarici*)

Drippy gill is a widespread mushroom disease but is rarely of economic importance. *Pseudomonas agarici* is a group III green-fluorescent pseudomonad closely related to, if not identical with, *P. cichorii*. Not all *P. agarici* isolates produce a green-fluorescent pigment on King's medium B but may do on pseudomonas agar F (6.2.7); some give a weak or perhaps even negative hypersensitive reaction in tobacco although usually not when freshly isolated.

Symptoms. The disease is characterized by small droplets of bacterial ooze which exude from small, dark, watersoaked lesions on the gills. The disease tends to have a patchy distribution within the flushes.

Isolation. Suspend droplets of ooze in sterile water, examine microscopically (1.3 and 1.4) and streak the suspension on plates of King's medium B (6.2.7) if bacteria are seen. Incubate and if the dominant colony form is raised or low convex, butyrous, round, 1–2 mm diameter in 2–3 days (typical of *P. agarici*) but appears non-fluorescent, re-streak such colonies on pseudomonas agar F (6.2.7).

Test procedure. Suspend 2–3 such typical green-fluorescent colonies in sterile water and from each streak a nutrient agar plate and inoculate a set of the following LOPAT media:

1 5% sucrose nutrient agar (6.2.17).
2 Pectate gel (6.2.12) or potato tuber slice (5.3.6).
3 Arginine medium (6.4.3).

Incubate and disregard results from any set(s) if the plates streaked from the same suspension are a mixture of colony types. After 2 days record the oxidase reaction (6.6.2) from the nutrient agar plate and use growth from it to make a hypersensitivity test in tobacco. Also record potato rot or degradation of pectate gel. After 3 days record arginine reaction. After a further 2–3 days record tobacco reaction.

Presumptive diagnosis. A presumptive diagnosis of drippy gill can be made if green-fluorescent bacteria with typical colony morphology and with the following characteristics are isolated from characteristic bacterial ooze from the gills of caps:

1 No levan on 5% SNA.
2 Oxidase positive.
3 No pectolytic activity.
4 No arginine dihydrolase activity.
5 Tobacco hypersensitivity positive (rarely negative).

Confirmation of diagnosis. Confirm the diagnosis if the bacteria presumptively diagnosed as *P. agarici* produce typical drippy gill symptoms in a host test on mushroom (5.3.8).

For further information on the disease see Young (1970).

3.7.3 Bacterial pit (*Bacillus polymyxa*)

Bacterial pit is a widespread mushroom disease and under certain conditions can be serious, although affected crops may still be marketable. The aetiology of the disease is not fully understood. Although *Bacillus polymyxa* is not always associated with the disease, it can incite pitting after artificial inoculation; the disease may possibly be associated with mite damage.

Symptoms. The disease is characterized by many small, discoloured, sunken pits in the cap.

Isolation and test procedures. Excise small pieces of pitted tissue, macerate in sterile water, examine microscopically (1.3 and 1.4) and streak suspensions in which bacteria are seen on nutrient agar and nutrient dextrose agar (6.2.10). Incubate and examine after 2–3 days for flat and usually irregular, spreading colonies on nutrient agar and heaped, mucoid colonies with a matt surface on nutrient dextrose agar which normally adhere to the agar surface. Prepare Gram stains from any such colonies and examine suspensions of water under phase contrast for spores. If spores (one/cell) are seen suspend 4–5 such colonies in 0.5 ml sterile water and pasteurize the suspensions (6.6.7).

After pasteurization spread the suspension on a nutrient dextrose plate, incubate and record growth and colony form. If spores are not seen under phase contrast streak characteristic colonies on soil extract agar (6.2.15) and re-examine for spores after 3 and 5 days; if present perform a pasteurization test (6.6.7).

Also spread the suspension on pectate gel (6.2.12) and inoculate O/F glucose medium (6.3.1) and nutrient broth; incubate the two former at standard temperature and the latter at 50°C in an accurate water bath. After 1–2 days record pectate liquefaction and after 3 days record O/F test and growth in nutrient broth. A presumptive diagnosis can be made at this stage.

Take young, freshly-picked mushroom caps. Wipe them with a bacterial suspension (10^6 colony-forming units/ml) (5.3.8). Use sterile water as a control. Incubate the caps in a humid chamber at 25°C. Observe for pitting.

Presumptive diagnosis. Make a presumptive diagnosis of bacterial pit if Gram-positive or Gram-negative but KOH-positive (6.1.3), straight, rod-shaped bacteria about 0.7 × 2.5 μm with the colony characters given above are isolated as the dominant organism from affected tissue from caps with typical symptoms of the disease.

The diagnosis is more certain if spores (one/cell) can be demonstrated and if the bacteria are shown to produce acid and gas in O/F media, to liquefy pectate gel and not to grow at 50°C.

Confirmation of diagnosis. Confirm the diagnosis if bacteria presumptively diagnosed as *B. polymyxa* produce pits in mushroom caps (5.3.8).

3.7.4 Mummy disease (cause uncertain)

Mummy disease is characterized by distorted caps which are usually stunted and tilted. Stipes may have swollen bases. The causal agent has not yet been determined although several pseudomonads and a rickettsia-like bacterium have been associated with the disease. The disease can spread very rapidly through mushroom beds but is generally considered to be relatively uncommon.

4

Recent Developments in Techniques for Rapid Diagnosis

4.1 Introduction

Many of the techniques used in diagnostic plant bacteriology derive from those developed in the medical field for human pathogens. In recent years there have been great advances in diagnostic techniques for human pathogens that have not yet been used routinely for plant bacteria, although some, and particularly those based on enzyme-linked immunosorbence (ELISA) and immuno electron microscopy, have been successfully adopted in plant virology. Some show considerable promise for plant bacteria. However, none have yet been developed sufficiently to be extensively used, and they cannot be applied to the diagnosis of more than the handful of specific plant bacterial diseases for which they have been studied to date without considerably more investigation. It seems probable that they will need to be specifically adapted or modified for use with each group of plant pathogens, and even in some cases for individual pathogens. It seems worth discussing these methods here, and giving details of some, if only to stimulate interest and further work.

4.2 The use of AP1 and similar systems

These rapid diagnostic kits, which are produced commercially, rely on the ability of a bacterial isolate in pure culture to produce particular enzymes under standard conditions. They have been designed for medical diagnosis and do not necessarily serve the needs of the plant bacteriologist. Kits range from those which detect the presence of enzymes within several hours to those which rely on some bacterial growth, usually over a 24-hour period. The AP1 20E and AP1 20NE perhaps have the greatest potential unless kits designed specifically for bacterial plant pathogens become commercially available; this seems unlikely, although Gardan *et al.* (1984) have made a preliminary study of kits for plant pathogenic pseudomonads with a view to possible commercial production.

The AP1 20E series is designed for the diagnosis of diseases caused by members of Enterobacteriaceae, which contains the plant pathogenic erwinias although they are not included in the diagnostic manuals or computer programmes which accompany the kits. The AP1 20NE kits are designed for a range of medical diseases caused by bacteria other than the Enterobacteriaceae including fluorescent and non-fluorescent pseudomonads. As yet little has been published on their use for diagnosis of bacterial plant pathogens and until this situation changes their use can only be in their evaluation as diagnostic tools, since the results they give for utilization of substrates or formation of products may differ from those obtained by more conventional means such as those described in Chapter 3. Nor can individual tests in kits be substituted for similar tests given in Chapters 3 and 6 of this book. There appears little doubt that many of the well-defined taxa in the genus *Erwinia* can be identified by the use of the AP1 20E system and its use (and that of other kits) for identification of *Erwinia* species is well documented by Mergaert *et al.* (1984).

Although the present code or pattern of results given for cultures applies only when tested as described in the instructions for use of the kit, it seems likely that incubation for plant pathogens needs to be at 30°C or even at 27°C for 48 hours, rather than the 37°C for 24 hours recommended for the growth of human pathogens and commensals since some plant bacteria will not grow, or grow only poorly at 37°C. Most *Erwinia* species give distinctive codes using the system, although this is not true for *Erwinia herbicola* and *Erwinia rhapontici* strains. As Mergaert *et al.* (1984) comment, this probably reflects the inadequacies of the present system of classification for these species. The use of the kit does not appear to distinguish between some pathovars and subspecies within the erwinias and *E. carotovora* subsp. *carotovora* and *atroseptica* cannot be differentiated.

Another system, the AP1 50 series, is available but at present its cost is high and a number of tests are apparently irrelevant for plant bacteria.

4.3 **Serological techniques**

4.3.1 Introduction

The avidity of the classes of protein known as antibodies which are produced by vertebrates for the large foreign molecules (antigens) which stimulate their production has been harnessed by medical diagnosticians for many years. The basic study of the immune response (immunology) and the study of sera (serology) are well documented by Roitt (1984) and that of their use in diagnosis by Weir (1978). The full potential of serological methods for diagnosis of bacterial diseases of plants has not been obtained and although much research has been done, the majority of workers involved with diagnosis rarely use serological methods. Primarily this is because, although antisera themselves are not particularly difficult to produce, antisera with the required specificity are rarely available and their development and production is laborious.

This book is not the place to discuss in detail the principles of *in vitro* interactions of antigen and antibody but a few important points must be briefly discussed. This should enable the reader to understand some of the basic technical terms used in relevant scientific papers and perhaps understand the principles behind the techniques used.

Antibodies. An antibody is a type of globulin protein that is formed during the immune response (i.e. an immunoglobin). It is basically Y-shaped, with two identical regions at the end of the arms that will bind specifically to the antigen that induced their production. An individual mammal may produce in excess of 10^6 different antibodies, each with a unique binding site that will bind a unique antigen. The remainder of the antibody molecule shows less variability and binds to cell and tissue components involved in different aspects of the immune response. It is this latter end to which fluorescent dyes or enzymes can be conjugated. Binding to this non-specific end of the molecule does not necessarily affect the antigen specificity of the other end.

Antibodies are produced by white blood cells (lymphocytes) in response to the presence of foreign proteins and other molecules recognized by the whole organism as not being 'self'. There are five major classes of antibody but only two of these are of relevance to serodiagnosis — immunoglobulins G and M (IgG and IgM). IgG is the most abundant immunoglobulin in the blood stream. IgM is a larger molecule than IgG, consisting of 5 Y-shaped molecules in a 'star' or 'crab' shape and the ten binding sites cause rapid precipitation of antigen found in the blood stream.

Not only can the animal body distinguish between one antigen and another, producing unique antibodies to each; it can also distinguish between self and non-self. Antibodies are not normally produced against self. An antigen (*anti*body *gen*erator) is then a molecule regarded as foreign which can elicit an immune response resulting in the production of antibodies which have a high degree of avidity for it.

Antigens. A bacterial cell contains many antigens — proteins, mucopolypeptides, polysaccharides. Some of these can be divided into smaller fragments by enzymes, and these may each elicit a specific and unique antibody. Thus an antigen can contain one or more antigenic determinants or epitopes.

When mixed together in the correct proportions soluble antigens and antibodies will combine and precipitate — the basis of precipitin reactions. Where antigens are on the surfaces of cells antibodies can attach to them and form bonds between the cells, causing them to clump or agglutinate — the basis of agglutinin reactions. However, if either antigen or antibody is in relative excess the reaction will be reduced. For any serological reaction it is therefore important to have antibodies and antigens in optimal proportions, and to obtain this we must react a range of antibody dilutions against a range of antigen dilutions. Further details of the basic principles of immunology and serology may be found in Roitt (1984), and Glynn and Steward (1977).

One of the major limitations of serodiagnostic tests is the ability of antibodies to react *in vitro* with antigens with slightly different structures. This can result in cross-reactions occurring in which a serum against one bacterium reacts with a different bacterium. Again, the variety of the antigenic determinants in a bacterial cell wall will often result in common antigens, antigens which are shared by more than one type of bacterium, which also cause cross-reactions. These can often be eliminated by the laborious process of cross-adsorption which removes common antigen once it has been serologically identified. This process may also reduce unwanted, non-specific reactions in which bacteria react by precipitation or agglutination with components of serum that are not antibodies.

For use in diagnosis, antibodies should always be used at dilutions close to their titres. Titre is often defined as the reciprocal of the highest dilution at which an unequivocal reaction takes place. For certain tests, e.g. the fluorescent antibody test, this definition may be narrowed to the reciprocal of the dilution at which the maximum reaction takes place. This will be discussed in the relevant section. When antiserum is used at this titre, the interference from cross-reactions will be minimized and the avidity for antigen is greatest.

The problem of cross-reactivity necessitates the strict use of controls in serological tests. These usually include the use of pre-immune (normal) sera and the use of a positive control homologous antigen — usually the strain of bacterium to which the antiserum was produced. Where non-specific protein – protein binding between antigen and antibodies is suspected, this may be reduced by the use of detergents or specific blocking agents, e.g. other proteins. Where pre-immune sera are used they must be taken from the same individual animal as the antiserum.

4.3.2 Production of antigens

Antisera may be produced against live or killed bacterial cells, or against bacterial products such as toxins or against specific cell fractions such as ribosomes or glycoproteins. The techniques described here are for the production of antisera against whole cells.

It is imperative that cell suspensions (the antigens) are free from constituents of the growth medium on which they were grown. Where possible cells should be cultured on a relatively simple medium such as glycerol agar. They should be washed several times by centrifuging at approx. 10 000 ***g*** for 15 minutes, resuspending the pellet in sterile, distilled water or sterile 0.01 mol phosphate-buffered saline at pH 7.2 (6.7.2). Cells still in the logarithmic phase of growth should be used and harvested from solid medium in sterile, distilled water or buffer. For more rapidly-growing bacteria 12–18 hour cultures are usually used; for slow growing bacteria, e.g. coryneforms, 4–7 day cultures may be needed.

For motile bacteria, flagella provide excellent antigens. Flagella antigens may be shared by closely-related bacteria however. These are often termed H antigens, whereas cell wall or somatic antigens are termed O antigens. To produce highly motile cultures, a medium other than the stock medium should be used. If the stock medium is nutrient agar, glycerol agar provides a useful medium for producing motile cultures. After preparation, live cell cultures may be preserved in phosphate-bufferred saline for a week at 4°C.

For bacteria-producing mucoid colonies, washing the cells may not remove all extracellular polysaccharide that can mask other antigens. It is possible to remove or fix such extracellular products by heating washed cells for 30 minutes in a boiling water bath or fixing them by dialysing against 2% glutaraldehyde (which should be used with care because of its reported carcinogenicity) for 12 hours at 4°C. Excess glutaraldehyde is removed by back dialysis with several changes of sterile phosphate-bufferred saline. These inactivated or dead preparations may be preserved by thiomersal (0.01%) or sodium azide (0.02%) and storage at 4°C. When required, the appropriate volume is removed, the bacteria centrifuged and washed several times and resuspended at the same concentration. When centrifugation may remove soluble antigen, back dialysis with several changes of sterile phosphate-buffered saline will suffice.

4.3.3 Production of antisera

Most antisera can be conveniently produced in rabbits,* which are among

*In some countries official licences are needed for antisera production.

the easiest of small mammals to rear and keep for antiserum production. Some plant pathogenic bacteria are toxic to rabbits if injected intravenously. In particular, some *Erwinia* spp. particularly *E. carotovora* or *E. chrysanthemi* can cause acute discomfort or death even when heat-killed (60°C for 30 minutes) suspensions are used. These reactions can usually be avoided by immunizing with a subcutaneous or intramuscular injection before using the intravenous route. The production of an antiserum with a high titre. (i.e. with large amounts of specific antibodies) depends upon several factors. The most important of these is the continued presence over a period of several weeks of the antigen in the blood of the rabbit. This can be achieved in several ways:

1 By multiple intravenous injections, increasing the amount of antigen with time. Injections at comparatively short intervals are required because the liver will filter out and destroy bacteria present in the blood.

2 By introducing the antigen into muscle or skin tissue. This will produce a slow release of antigen into the blood stream ensuring a continued presence over a period of several days. This process can be made more efficient by emulsifying the antigen with an oil known as an adjuvant. This will produce an even slower release of antigen into the blood stream.

The most specific antisera are usually produced by short immunization procedures. Continued immunizations may well increase titre but may also increase the amount of common antigen. The intravenous route is generally considered best for antisera to whole cell preparations. Most antibodies produced by the above methods will be as the result of a humoral response and will consist of the two major classes of immunoglobulin, IgG and IgM. IgM is well suited to precipitin reactions and IgG is well suited to non-precipitin reactions such as ELISA and immunofluorescent antibody techniques.

Antibody production can be stimulated if antigens (e.g. whole bacteria cells) are first emulsified with an adjuvant before inoculation by the intramuscular or subcutaneous route. The most commonly-used adjuvants are Freund's complete adjuvant and Freund's incomplete adjuvant.

Immunization schedules that have proved effective for production of antisera against plant bacteria (other than coryneforms) in our laboratory include those shown in Table 4.1. Most of the Gram-negative plant pathogenic bacteria should produce a titre in excess of 1:1600 by these methods.

Coryneform bacteria often produce lower titres. Antibodies produced to plant pathogenic coryneforms usually cross-react with many other coryneform bacteria often found in soil and on the surfaces of or within healthy plants. The following method has been found to produce antisera

Table 4.1. Immunization schedules for the production of antisera against plant bacteria

Day	Route	Volume (ml)	Antigen
METHOD 1			
0	Subcutaneous	1.0	18–24 hour slope culture in 4 ml phosphate buffered saline (days 0–17)
7	Intravenous	0.2	
10	Intravenous	0.5	
14	Intravenous	1.0	
17	Intravenous	1.0	
21	Test bleed. If the titre is more than 1:1600 bleed from the ear vein or by cardiac puncture. If less than 1:1600 proceed as follows:		
21	Intravenous	2.0	18–24 hour slope culture in 4 ml phosphate-buffered saline
24	Bleed from ear vein or by cardiac puncture. If a high enough titre cannot be obtained start again with another rabbit.		
METHOD 2			
0	Intramuscular	1.0 ml in each hind leg	18–24 hour slope in 4 ml phosphate-buffered saline emulsified with an equal volume of Freund's Complete Adjuvent
21	Intravenous	1.0	18–24 hour slope culture in 4 ml phosphate buffered saline (days 21–24)
24	Intravenous	2.0	
28	Test bleed and if titre is more than 1:1600 bleed from vein or by cardiac puncture. If titre is not high enough inject a further 2 ml intravenously and test bleed at 31 days, start again with another rabbit if the titre is still too low.		
METHOD 3			
0	Intramuscular	1.0 ml in each hind leg	18–24 hour slope culture in 4 ml phosphate buffered saline (days 0–14)
14	Intramuscular	1.0 ml in each hind leg	
21	Test bleed as before		

with titres more suitable for diagnostic use. Because many coryneform species are slow-growing, populations of 10^8-10^9 cells of a 3–7 day culture should be used.

Injection times	Route	Volume	Antigen
Weekly intervals for up to seven weeks	Intramuscular	1.0 ml in each hind leg	10^8-10^9 cells/ml of 3–7 day slope culture in phosphate-buffered saline emulsified with an equal volume of Freund's Incomplete Adjuvant

Test bleed after four or five weeks. Titres should exceed 1:320.

It should be noted that long immunization schedules tend to decrease antibody specificity.

4.3.3.1 Methods of injection

Subcutaneous route. Lay the rabbit on its side and wet the flanks so that the fur may be parted and the skin exposed. Swab with alcohol. Distribute half of the inoculum into 10–20 'blisters' made by injecting droplets just under the skin. Distribute the other half by the same method in the other flank.

Intramuscular route. Place the rabbit in a cage suited for the task or support in a towel used as a sling, to reduce movement. Pull out a hind leg, part the fur to reveal the rear of the upper thigh muscle. Swab with alcohol and inject to a depth of approximately 1 cm. Deliver the required volume of antigen. Repeat with the other leg.

Where adjuvant is used, mix with an equal volume of antigen and either mix on a Vortex blender or pass from one syringe barrel to another via an adapted needle with Luerlock sockets at each end, until emulsification is complete. This can be detected as a stiffening in the emulsion, but should be tested by pipetting a droplet on to the surface of water. If emulsification is complete the droplet will float intact; if not complete, spreading will occur.

Intravenous route. Cage the rabbit as above. Locate the marginal ear vein and shave the hair to expose it. Swab with alcohol and introduce the needle into the vein towards the base of the ear.

With all injection methods, the sites of inoculation are prone to inflammation, swelling and infection. Rabbits should be inspected regularly and any ulcerations or abscesses treated immediately.

4.3.3.2 Bleeding

Place the rabbit in a bleeding cage or have an assistant hold it firmly on the bench facing you. Remove the down from the back of lower ear margin with a sharp scalpel. Swab the area around the base of the ear and using a scalpel make a nick in the marginal vein. The cut should be longitudinal — along the vein; *it should not be made across it.*

Trap the vein below the cut between thumb and forefinger. Warm the ear tip with a cloth soaked in warm water to start the blood flow if necessary, and collect the drops of blood in a centrifuge tube. Collect approximately 5 ml of blood.

Allow the blood to coagulate after scoring the edge of the meniscus with a toothpick to free the clot from the sides of the collecting vessel. Decant. Coagulation should be complete after 4 hours at 37°C. Samples may be left overnight at 4°C to ensure that the process is complete. Spin the supernatent twice at 4000 g for 10 minutes to remove the excess red blood cells. Determine the titre. If high enough take another 20 ml of blood.

Further bleeds of 20–30 ml may be taken at two-weekly intervals for as long as the titre remains high enough, although later bleeds may produce less specific antisera. Bleeding by cardiac puncture requires confidence and some skill. It is best not attempted by the inexperienced except under supervision by and instruction from someone experienced in the technique.

4.3.3.3 Normal serum (pre-immune serum)

Non-specific binding may occur with plasma proteins other than specific antibodies. 20 ml of normal serum should therefore be removed before immunization begins for later use as a control in *in vitro* tests for antibody–antigen reactions.

4.3.3.4 Purification of antisera by ammonium sulphate precipitation

Dilute antiserum ten-fold in distilled water. Add an equal volume of saturated ammonium sulphate (pH 7.0) dropwise with gentle agitation. Leave overnight at 4°C or for several hours at ambient temperature, with gentle agitation to allow the globulins (antibody proteins) to precipitate. Remove by centrifuging at 8000 g for 10 minutes. Dissolve the pellet in the original volume of 0.1 mol TRIS buffer pH 7.4 (6.7.5) or 0.05 mol phosphate-buffered saline pH 7.2 (6.7.2) diluted 1:1 in distilled water. Dialyse against 1 litre of the same buffer for 1 hour at room temperature to remove excess sulphate ions.

Such preparations are often termed partially purified. Antibody concentration can be increased by suspension in lower volumes of buffer. Protein content can be determined approximately by reading at 240 mm in a spectrophotometer. (1.4 absorbance units = approximately 1 mg/ml).

4.3.4 Fluorescent antibody (FA) or immunofluorescent (IF) techniques

4.3.4.1 Introduction

These techniques probably offer the most accurate and useful serodiagnostic tests for the bacteriologist, since the reaction can be visualized under the microscope and cross-reacting antigens often recognized. They are based on the ability of a fluorescent compound to bind to antibody. When this conjugated antibody binds to bacteria, its presence can be visualized in a microscope using an ultraviolet light source and suitable filters for the fluorescent dye used. Cells that combine with antiserum appear with a fluorescent halo around the cell wall.

There are two methods by which these processes can be achieved: direct and indirect antibody techniques. Direct techniques use the fluorescent compound conjugated directly to the antibodies produced against the bacterium. The intensity of fluorescence may often be limited and each antiserum must be conjugated.

Indirect techniques use a technique in which the bacterium is coated with antibodies produced in a rabbit. The coated bacterium is then reacted with antibodies produced in sheep or goats against rabbit immunoglobulins of the IgG type — themselves antibodies. The antibodies from sheep or goat are conjugated with a fluorescent dye. These conjugated sheep or goat antirabbit antibodies will bind to any rabbit immunoglobulins. When used in the FA technique, they bind to any antibody attached to the bacterial cell wall allowing the cell to be seen with a fluorescent microscope. Since one conjugate can be used for all bacterial antisera produced in rabbits, the indirect method has the great advantage that each antisera used by the diagnostician does not have to be separately conjugated, as in the direct method.

The fluorescent compound usually used is fluorescein isothiocyanate (FITC) which, under the correct wavelength of ultraviolet light, fluoresces bright green.

The FITC conjugated sheep or goat antirabbit IgG may be obtained commercially from suppliers to medical laboratories.

4.3.4.2 Production of reagents and test method

To obtain a specific reaction with crisp fluorescence, both antibody preparations must be used close to their titres for the bacterium being used. These values will be referred to as working dilutions. It is always best, therefore, to determine these values for each antiserum used. It is also advisable to use fairly low cell concentrations of approximately 10^6 cells/ml, to avoid antigen excess, or preferably test against a range of cell concentrations, and use that which gives the highest titre. Plastic-coated multi-window microscope slides* should be used to determine titres or 7 mm diameter windows may be etched on the slide with a diamond marker.

Pipette 25 μl of a 10^6 cells/ml suspension of the bacterium to which the antiserum was produced to each window of two slides and allow to air dry. Heat-fix gently in a Bunsen flame. Prepare a range of eleven antiserum and normal serum dilutions serially two-fold from 1:10 to 1:10240 in sterile 0.01 mol phosphate-buffered saline (PBS) at pH 7.2 (6.7.2). To one slide, pipette 25 μl of each antiserum dilution to a window. Add 25 μl of PBS to the last window as an FITC conjugate control. To the other slide pipette 25 μl of each dilution of normal (pre-immune) serum in the same manner. Place the slides on damp blotting paper and cover to maintain a dark, moist atmosphere. Incubate at ambient temperature for 30 minutes. Shake the droplets off and rinse with PBS. Wash three times for 5 minutes in PBS and gently blot dry. Pipette 25 μl of an appropriate dilution (1:50–1:400 for the Sigma product) goat anti-rabbit gamma globulin-FITC conjugate† to each window. Incubate as before for 30 minutes. Rinse and wash as before. Gently blot dry. Pipette approximately 10 μl of 0.1 mol phosphate-buffered glycerine pH 7.6 (6.7.4) to each window and cover with a coverglass.

Observe at high power, preferably under oil immersion lens ($\times$ 60 or $\times$ 100) starting with the FITC control. If this control shows no fluorescence, proceed to the normal serum controls. If these show no fluorescence proceed to the antiserum dilutions starting with 1:10. Low dilutions may give poor staining due to antibody excess and may also show background fluorescence seen as a dull-green background. The intensity of cell wall fluorescence should increase with increasing dilution and the background fluorescence should decrease. At one dilution the intensity of fluorescence will fade, often dramatically, and subsequent dilutions will show even less fluorescence although individual cells may still be detected.

*Obtainable from C.A. Hendley Ltd, Oakwood Hill Industrial Estate, Loughton, Essex, UK.

†Available from Sigma Chemical Co., Fancy Road, Poole, Dorset BH17 7NH, UK (Product No. F6005).

The titre for FA work is determined as the highest dilution at which optimum fluorescence is observed.

The titre of the FITC conjugate should then be determined by running a series of conjugate dilutions against cells treated with antiserum at its titre. Titres may be used as working dilutions but for routine screening a working dilution of one half to one quarter of the titre may be used to allow for antigenic differences in the cell walls of different strains of the bacterium. Occasionally a different strain or culture may actually give a higher antiserum titre than the homologous antigen.

For diagnosis it is wise to use a two-fold dilution series of antiserum from 1:10–1:5000. Normal (pre-immune) serum may be used at a single dilution, e.g. 1:20 or 1:40. A dilution series of bacteria (e.g. 10^5-10^8 cells/ml) and of macerated plant tissues (neat, 1:10, 1:100, 1:1000) should also be incorporated as shown in Fig. 4.1. A positive control slide using homologous antigen, i.e. known bacterium is essential. The method used should be as for titre determination. Positive controls must always be washed in a separate bath, to avoid contamination.

4.3.4.3 **Interpretation of results**

If either normal serum or FITC controls show cell wall fluorescence then the test should be repeated using normal serum at all dilutions to determine the extent of non-specific binding. This is rare, but when it occurs the antiserum or FITC conjugate may have to be discarded.

If almost all of the cells of the test bacterium fluoresce brightly at the titre or the working dilution then a positive presumptive diagnosis may be made. Clear, cell-wall fluorescence at dilutions of a half or a quarter of the titre can also be regarded as positive. End points below this must be regarded as suspect. It is imperative that comparison with the positive control is made, as subtle differences in the quality of fluorescence must be looked for. These include: patchy or incomplete cell wall fluorescence, indicating some common antigens; weak fluorescence when the intensity of cell wall fluorescence is much weaker than the positive control, or is extremely 'thin' and the cells only just distinguishable — weak fluorescence may be due to the presence of common antigens, inadequate washing or use of the FITC conjugate at too low a dilution, or antigen/antibody excess; a very low percentage of cells fluorescing which occurs with antigenically diverse cell populations is particularly common with the Gram-positive coryneform bacteria or with old cultures.

Since non-fluorescing cells will not be seen under ultraviolet light it is not possible to determine how many cells are fluorescing directly. If phase contrast as well as ultraviolet illumination is available on the same microscope a quick change from UV to phased lighting will enable

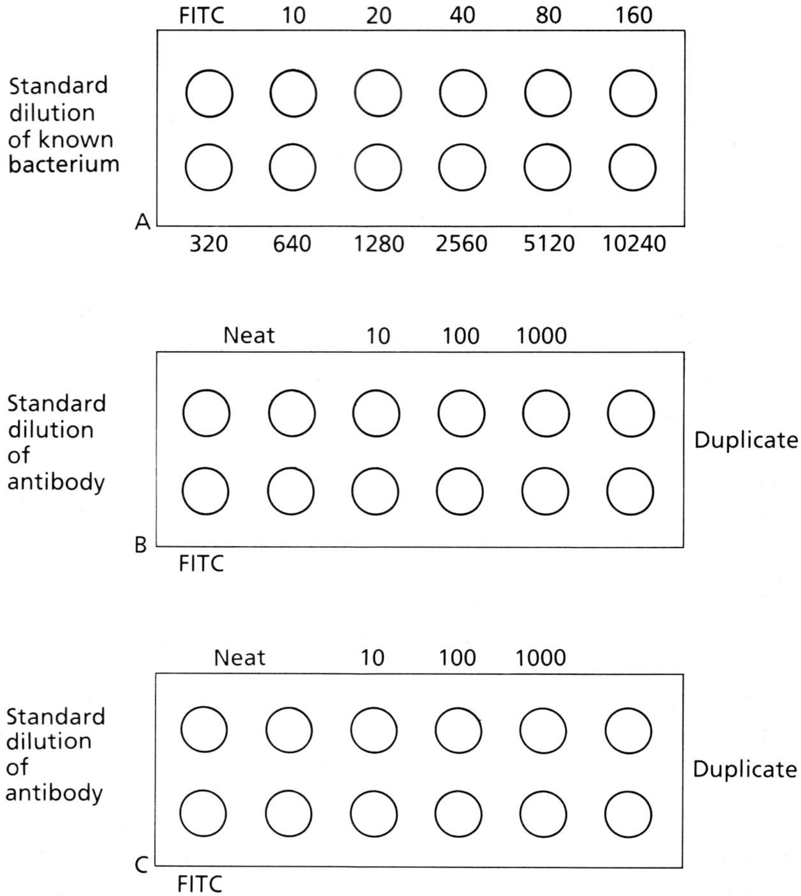

Fig. 4.1. Dilutions of:
A Antiserum.
B A known bacterium.
C A dilution of macerated plant tissue.
Dilutions increase from left to right.

an estimate of the proportion of cells fluorescing. Otherwise, a stained preparation of the same cell suspension can be compared with the FA preparation.

4.3.5 Slide agglutination

This is the most simple of serodiagnostic tests. A positive reaction is given by agglutination of bacterial cells due to multiple bridging between cells by specific antibodies attached to antigens on the cell walls and on flagella. The reaction takes place in a drop of bacterial suspension on a microscope slide and is seen as a clumping of cells. Gross agglutination may be seen by eye but weak agglutination is best seen under a low-power stereomicroscope.

An antiserum for slide agglutination work should be tested with the bacterial culture at *c.*10^9 cells/ml against which it was prepared (the homologous antigen) to determine the range of doubling dilutions at which good, easily determined agglutination occurs. If possible it is preferable to test other isolates of the same bacterium as well, to determine the likely dilution range for optimum agglutination for most isolates of the bacterium. From the tests choose a working dilution for the antiserum that is two dilution steps lower than the least dilution at which optimum agglutination occurs. Thus if the optimum occurs at 1:8, 1:16, 1:32, 1:64 and 1:128 take a working dilution for the antiserum of 1:32. A turbid suspension of *c.*10^9 cells/ml of homologous bacterium should be prepared and preserved in 0.01% thiomersal or 0.02% sodium azide and stored at 2–4°C for future use.

4.3.5.1 **Test method**

Suspend a colony of the unknown bacterium from an isolation plate, or from any solid medium in sterile 0.05 mol phosphate-buffered saline (pH 7.2) (6.7.1) to give a turbid suspension (*c.*10^9 cells/ml). Place a drop of normal serum, a drop of appropriate antiserum at half its working dilution and two drops at working dilution from left to right in a row along a microscope slide. Below each of the three left hand drops place a drop of the unknown bacterial suspension and below the right a drop of the homologous bacterial suspension. Draw a sterile loop between each drop of suspension and the drop of serum below it to join them and then rock the slide gently about its long axis to mix the antigen and antiserum. Place on a damp filter paper in a Petri dish and record agglutination after about 1, 2 and 5 minutes, checking with a good hand lens or a low power stereomicroscope for weak agglutination.

A positive result is agglutination, comparable with that between homologous suspension and its antiserum, by the unknown suspension against antiserum at working or half-working dilution and no agglutination against normal serum in a standard time (of less than 5 minutes) which should be determined for each bacterium. Generally speaking, incubation times should be minimal to avoid cross-reactions with heterologous bacteria. Due to the high bacterial populations needed for reaction to be visualized, the slide agglutination test (SAT) cannot really be used to detect bacteria in plant tissues. One of the most useful applications of the SAT is for identifying cultures of *Erwinia amylovora*, strains of which, so far as is known, are antigenically homogenous and distinct in SAT test from other bacteria found in association with *E. amylovora*.

4.3.6 Latex agglutination

This test, unlike slide agglutination, will give a reaction in the presence of comparatively low numbers of cells of the homologous bacterium. It therefore has considerable promise for detecting and diagnosing pathogens in plant tissue directly. It uses small latex beads* to which the antibodies in the antiserum are adsorbed. A positive reaction is seen as clumping of the beads. For this test, antisera should be purified either by ammonium sulphate precipitation (4.3.3.4) or by a method suitable for purifying IgG such as affinity chromatography. The antiserum dilution used in the conjugation to the beads is critical, and conjugations with a range of antiserum dilutions should be prepared. The product is usually stable for several months. The test is rapid, but cross-reacting bacteria will cause agglutination; plant sap may also cause non-specific agglutination often indistinguishable from true agglutination. The method given below has been designed to minimize such problems but cannot guarantee to remove them.

4.3.6.1 Preparation of antibody sensitized latex particles

Prepare an appropriate dilution series of the purified or partially purified antiserum in 0.1 mol TRIS buffer pH 7.4. (6.7.5). To 5 ml of each dilution add an equal volume of polystyrene latex beads (8 μm diameter) suspended 1:15 (v/v) in 0.85% sodium chloride solution.

Shake vigorously and allow to stand for 1 hour at room temperature with occasional shaking.

Centrifuge at 4000 g for 30 minutes to pellet the latex beads. Remove supernatent carefully. Stabilize the sensitized latex particles by washing twice in 0.02% polyvinyl pyrrolidone in 0.85% sodium chloride. Collect the latex by spinning at 4000 g for 20 minutes. Resuspend in 5 ml 0.01 mol TRIS buffer pH 7.4 containing 0.02% sodium azide and 0.004% bovine serum albumin (Sigma Chemical Co.). Filter through Whatman No. 2 filter paper under vacuum. Store at 4°C.

4.3.6.2 Determination of titre

Take up some sensitized latex (4.3.6.1) into a glass capillary tube by touching the surface of the liquid. Take up a similar volume of antigen (10^6 cell/ml suspension of bacteria). Repeat for all dilutions of antiserum with sensitized latex.

Fix the tubes radially to a vertical plate (like the spokes of a wheel) and

*Obtainable from Sigma Chemical Co. or Difco Laboratories.

rotate the plate at about 4 revs/minute with an electric motor in order to mix the latex and bacteria for up to 1 hour. Examine for visual clumping by eye and check for weak agglutination using a dissecting stereoscopic microscope or, if available, phase contrast or dark field illumination. Agglutination normally occurs within 5–10 minutes. Extending the mixing period may increase the sensitivity of the test but may also decrease the specificity.

4.3.6.3 **Test procedure**

Macerate lesions that contain bacteria in sterile water, decant off from the layer of plant debris and use the supernatent as antigen against a dilution range of antisera containing sensitized latex. Use homologous antigen at 10^6 cells/ml as one control and healthy plant sap as another, both against the same dilution range of antiserum with latex.

Agglutination may occur within several minutes, but obvious agglutination in the presence of small numbers of bacteria may take up to an hour to occur. A positive diagnosis can be made if the bacteria in the suspension of infected plant tissue agglutinate to within 50% of titre and there is no agglutination with the healthy plant sap or, if there is, that it can be differentiated microscopically from the bacterial agglutination in infected tissue. Plant sap, especially from the woody stems of orchard trees and hardy ornamentals, can cause non-specific agglutination which obscures observation of bacterial agglutination.

4.3.7 Immunodiffusion (ID)

This is a procedure in which soluble antibodies and/or antigens diffuse independently through a gel, usually of agar. Where the homologous antibodies and antigens meet in equivalent proportions, a precipitin reaction occurs which can be seen as the formation of a white precipitate. A wide range of methods are available. One of the most widely used is the Ouchterlony double diffusion test where antigen and antibody are both actively diffusing through the gel. This is the technique described below.

4.3.7.1 **Gel preparation**

Use a purified agar, e.g. Difco purified agar. Prepare a 0.8% agar suspension containing 0.85% sodium chloride and 0.05% sodium azide as a preservative. Pour plates (15 ml agar per Petri dish), or use sufficient to flood a glass microscope slide, on a flat surface. Values given below are suitable for use with Petri plates. Prior to use cut six peripheral wells equidistant from a central well, preferably with a template. For Petri dishes, all wells should be 5 mm diameter, peripheral wells 4 mm apart and approximately the same distance from the central well. Add one small drop of molten agar to the base of each well to seal the well. (This is not essential with plastic Petri dishes.) Agar plugs should be removed by a suction pipette.

Add 50 μl of undiluted antiserum to the central well. Add 50 μl of known control (homologous) antigen to two opposite peripheral wells. The remaining four can be used to test unknown cultures. Cover the plates and incubate in a humid chamber for 24 hours.

4.3.7.2 Culture preparation

Grow the test bacteria on a suitable agar medium and suspend in a 0.1N sodium hydroxide solution. Centrifuge to remove cells, autoclave the supernatent and store at 4°C. Sodium azide (0.05%) may be added as a preservative. Direct diffusion of plant tissue extracts may be performed but non-specific bands may form.

4.3.7.3 Interpretation of results

Look for one or more white bands between the central well and peripheral reference wells. If the same band pattern is found for the test wells, homologous antigen is indicated. Each band corresponds to an antigen-antibody complex forming at optimum concentration. Spurs may be seen between adjacent antigen wells as seen in Fig. 4.2. These indicate that the antigens are not identical but are related. This phenomenon can be used to determine antigenic similarity and is thus important in diagnosis by comparison with known strains. The positioning of antigens in adjacent wells in various combinations will increase the diagnostic value of the test. For further details see Van Regenmortel (1982) and Weir (1978).

4.3.8 ELISA

Whereas enzyme-linked immunosorbent techniques (ELISA) are used routinely to detect and diagnose diseases caused by viruses, their use in plant bacteriology is still in its infancy. ELISA uses clear polystyrene plates usually containing eight rows of twelve wells of approximately 400 μl volumes. The most commonly-used technique in plant disease diagnosis is the double antibody sandwich technique in which specific antibody (usually purified IgG or IgG partially purified by ammonium sulphate precipitation) is conjugated to the polystyrene wells. A suspension of the test bacterium or homogenized plant tissue or sap suspected of containing it is placed in the well and allowed to react with the antibodies. After washing, more specific antibody conjugated to a suitable

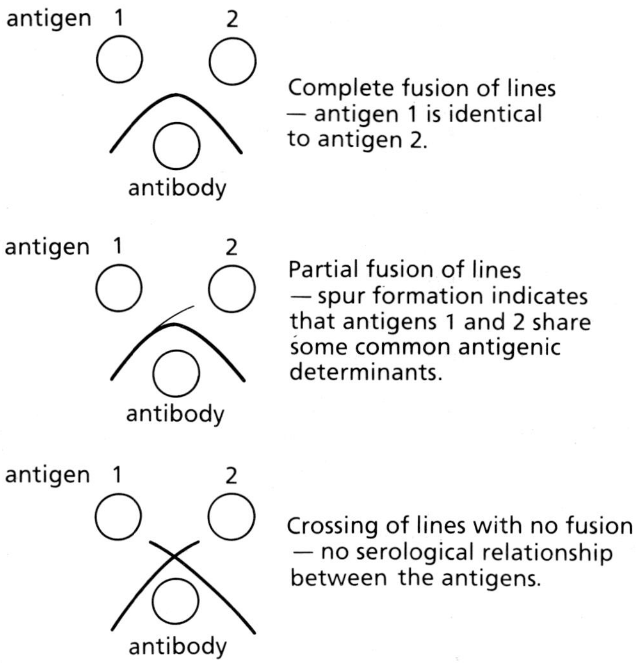

Fig. 4.2. Basic precipitation patterns in double immunodiffusion.

enzyme, e.g. alkaline phosphatase is placed in the wells and allowed to combine with any bacteria which have adhered to the antibody coating on the walls of the cells. After washing again, the appropriate substrate for the enzyme is added to the wells and the colour change produced can be detected by eye or measured colorimetrically. The system can be semi-automated by the use of a scanning colorimeter so that each well on the plate can be assayed individually within a very short time. The quantity of antigen present is proportional to the absorbance at a specific wavelength.

Several other ELISA techniques are available but are not discussed here. The major advantages of ELISA are that large numbers of samples can be assayed quickly and the system can be semi-automated. It can be extremely sensitive and therefore can use smaller quantities of antisera than other serological tests. The method is also quantitative. Although used extensively for plant viruses, the disadvantages in its application to diagnostic bacteriology appear to negate the advantages at present and so far relatively little has been published on the method. Bacteria are much more antigenically diverse than viruses; cell walls possess many different antigenic determinants many of which are common to other bacteria. Bacteria are much larger than viruses and the application of techniques successful for diagnosis of viral diseases may not be applicable to bacteria. More regard is needed in the method for fixing bacteria to the wells. Cross reactions cannot be distinguished from normal reactions, and for this reason fluorescent antibody stains are often preferred. Clark and Adams (1977), Clark (1981), Torrance and Jones (1981), Barzic and Trigalet (1982) should be consulted for further information.

4.4 Polyacrylimide gel electrophoresis of whole cell proteins (PAGE)

This technique is used extensively for the identification of many bacteria although there is little published for its use with plant pathogenic bacteria. Bacterial cells are treated chemically or physically to disrupt cell walls. The proteins are extracted and separated on a slab of polyacrylimide gel by electrophoresis under standard conditions for a standard time. The gel is stained for proteins. The proteins separate as bands across the gel and the pattern is characteristic for a given species. The complexity of these bands in the gel often requires analysis other than by eye. Perhaps the best method for analysis uses a laser densitometer to produce a graphical representation of the bands which can then be analysed by computer and compared with a library of known standards.

4.5 Fatty acid composition of whole cells

Bacterial cells contain many different fatty acids including straight chain, hydroxy, saturated, unsaturated and branched acids. The types and relative proportions of these are often unique to a particular taxon. Thus extraction and analysis of these, usually after esterification to the methyl esters can be used for identifying bacteria. Since as many as 50 different acids may be present in a cell, the method has potential for carrying identification to pathovar level and there is evidence that this can be achieved for at least some bacterial plant pathogens (Sasser and Miller, 1984).

The technique requires sophisticated chromatographic apparatus and facilities for computer analysis. Suitable software is now available for plant pathogenic bacteria but is expensive. Nevertheless this type of diagnostic technique holds much promise for the future.

Bacterial cells are harvested from agar plates, suspended in methanol and sodium hydroxide in tubes which are then sealed and heated. The cell walls are broken down in acid, the esters extracted in a suitable solvent which is separated from the aqueous phase, dried and the esters separated by gas liquid chromatography. A library of the profiles from known strains is collected and test bacteria compared with these.

4.6 Copy DNA (cDNA) probes

Bacteria possess identifiable phenotypic properties which can be used to classify them or identify them. These properties are based upon gene sequences in the DNA of the bacterial chromosome. All members of a single taxon possess a unique DNA sequence or sequences, which, if identified, can be used in diagnosis. The most obvious means of identifying a sequence is to prepare a length of DNA with a complementary base pair sequence commonly referred to as a probe. If this contains a marker, e.g. radioactively labelled phosphorus, successful hybridization, usually carried out at high temperature (*c*.80°C), of the DNA of the pathogen and the probe can be detected. These techniques are still in their infancy although probes have been developed for some plant pathogenic viruses.

Once probes have been produced the diagnostic method is relatively simple. The suspect plant sample is treated chemically to release any DNA present in the cells. This can be purified by gel electrophoresis and then 'blotted' on to nitrocellulose sheets which bind to the DNA present. The copy DNA probe is then added and the sheet baked at a suitable temperature. The probe does not bind to the nitrocellulose and so all that does not hybridize with the bound DNA can be washed off. Detection of radioactivity in the 'blot' implies successful hybridization and thus the presence of the pathogen. The disadvantages with the present systems are that laboratories must be able to deal with short-lived radioactive probes. Since tritium or ^{32}P labelled probes are short-lived they must be produced regularly. This can be facilitated by first producing a non-radioactive probe which can be stored until required. Before use, a radioactive label can be added by a process known as nick translation.

Non-radioactive probes have been developed and will undoubtedly become available in the next few years which, with further development of techniques, should simplify the diagnostic procedures so that nucleic acids released can be directly immobilized on the nitrocellulose or other solid support.

This type of diagnostic test probably offers the most sensitive and selective rapid diagnostic test procedure available for bacterial diseases of plants.

5

Host Tests

5.1 Introduction

The aim of a host test is to determine whether or not a suspected pathogen can cause disease symptoms in the host from which it was isolated under the environmental conditions in which infection and disease expression occur naturally. For many diseases this is impractical because even when these conditions are known with any precision it is difficult and/or costly to reproduce them. However, if certain criteria are fulfilled the host test can be a very good indicator of the pathogenic properties of isolated bacteria and the most precise method of determining their identity. In fact, for many plant diseases this test is the only method of confirming or negating an initial presumptive diagnosis.

The criteria to be followed can be summarized as:

1 Introducing the pathogen into or on to the plant in a manner which mimics the natural infection as closely as possible.

2 Introducing the pathogen into or on to the plant in cell concentrations similar to those that cause natural infections or at least do not exceed natural inoculum levels.

3 Cultivating the pathogen prior to inoculation under conditions which do not seriously affect virulence.

4 Cultivating and using plants which are naturally susceptible to infection and in conditions in which infection can occur naturally.

5 Incubating infected plants under conditions that could occur naturally.

6 Determining what can be regarded as typical symptoms. Many plant pathogenic bacteria will induce a hypersensitive reaction — an atypical symptom — if introduced in large numbers into non-susceptible hosts; this latter reaction, however, can provide an extremely useful presumptive diagnostic technique.

5.2 Hypersensitivity tests

Most bacterial plant pathogens can induce a hypersensitive response when injected into the tissues of a non-susceptible host plant. The plant cells immediately around such bacteria appear to be sensitized and killed by contact with the pathogen, and as a result, the extent of infection is limited, presumably because the pathogen cannot grow or is killed in the presence of dead cells. Pathogens in susceptible hosts seem to have evolved mechanisms to overcome the triggering of this reaction. Non-pathogenic bacteria and some pathogens, particularly those that are opportunistic or gall-forming, do not incite this hypersensitive response.

A variety of plants may be used. Plants with soft, succulent leaves are preferable and for many bacteria tobacco is the preferred plant since it is easy to cultivate and maintain, its large cavities beside leaf veins make it relatively easy to infiltrate inoculum and its reaction to many pathogens is well known. It is, however, a somewhat unreliable test plant for xanthomonads; other plants, particularly tomato, are preferred for many pathovars of the genus. For the reaction to involve enough plant cells to be seen, bacteria must be introduced in large numbers (10^8-10^9 cells/ml (5.4)) into the leaf tissue.

5.2.1 Tobacco hypersensitivity test

Use large-leaved, rapidly grown plants. White Burley is a commonly-used cultivar. Where possible, grow bacteria on nutrient agar. Adjust an aqueous suspension (5.4) from a 24–48 hour culture to a cell density equivalent to 10^8-10^9 colony-forming units/ml. This is usually equivalent to an optical density of approximately 0.3–0.4 at 600 nm.

Inject the mesophyll of the leaf lamina with the suspension by inserting the needle of hypodermic syringe into the cavity which runs along the side of the lateral veins. It is necessary to use a narrow gauge needle of approximately 0.6 mm external diameter. The diagonal of the needle aperture should be adjacent and parallel to the surface of the tissue.

Inject sufficient inoculum to flood the intercellular spaces of the mesophyll — apparent by water soaking of the tissue — but not at such high pressure that blisters appear on the leaf surface. Inject a separate area of the leaf lamina with sterile, distilled water as a control and with a culture of a pathogen from the same group as the culture being tested and known to cause a typical HR, as a positive control. Label the areas of the leaf lamina by wounding the leaf in a specific manner, e.g. by pin pricks or by using adhesive labels.

Inject more than one leaf and one sector of a leaf/suspension and arrange control and test injections on opposite sides of the main vein. For *Pseudomonas*, *Erwinia* spp., or for coryneform bacteria incubate plants in a well-shaded and ventilated glasshouse at a temperature of less than 30°C. False reactions may be obtained at higher temperatures. A typical hypersensitive reaction is shown in Fig. 5.1.

If growth cabinets are available incubate at 25°C and 85% relative humidity (RH), with a diurnal daylight regime of 16 hours.

For *Xanthomonas* species, use tomato or pepper plants (5.2.2) since not all *Xanthomonas campestris* pathovars induce a hypersensitive reaction in tobacco and those that do give reliable results only if plants are incubated at 16°C for 4 days prior to inoculation and at 33°C after inoculation.

A positive hypersensitive reaction is given by a rapid collapse and water soaking of inoculated tissue in 24 hours or at most 48 hours followed by a dry, light-brown necrosis of the watersoaked tissue within 3 days. Ideally conditions of the test should be such that a clear reaction is given in 24 hours. Yellowing or browning without collapse is not a positive reaction. Tobacco pathogens give a pathogenic reaction that is slower; the necrosis is usually much darker in colour.

Fig. 5.1. Hypersensitive reaction in tobacco. Crown copyright. Used by permission of the Controller of HMSO.

5.2.2 Tomato or pepper hypersensitivity test.

Proceed as for 5.2.1 but incubate plants at 28°C–30°C or for xanthomonads, at the pre- and post-inoculation temperatures given there.

5.3 **Pathogenicity tests**

5.3.1 Introduction

It is essential to note that many bacterial plant pathogens will sometimes elicit a disease reaction in many non-hosts if the bacterial population is high, e.g. $>10^7$ cells/ml or if the environmental conditions are adverse to the growth of the plant. Usually the symptoms created are atypical and do not occur with inoculations of lower cell numbers which do elicit a disease response in a truly compatible host. Nevertheless, these atypical symptoms are often characteristic for a particular non-compatible host and pathogen and can be used as diagnostic aids. For example, *Pseudomonas syringae* pv. *tabaci* causes angular leaf spots on young *Phaseolus* leaves; *Pseudomonas syringae* pv. *syringae* causes a distortion of young *Phaseolus* leaves. Thus such host tests could be used to confirm the diagnosis of a group I fluorescent pseudomonad (*Pseudomonas syringae*). Pathovar indentification can only be made on the normal host. However, it must be remembered that most phytopathogenic pseudomonads and xanthomonads may infect several closely-related host species. Some other pathogens such as *Erwinia carotovora*, *E. chrysanthemi*, *Pseudomonas viridiflava*, *P. marginalis*, *P. solanacearum* have relatively wide host ranges although host specific strains may exist as in *P. solanacearum*.

Some metabolic products of bacteria growth in or on culture media can produce reactions in test plants that can be confusing. To avoid this use suspensions of young bacterial growth, washed rapidly off agar media, as inoculum (5.4).

In host tests to confirm the identity of a suspected pathogen, a known, virulent culture of that pathogen should also be inoculated as a positive control and sterile water inoculated as a negative control.

5.3.2 Host tests on detached fruits.

Do not use old or stale material. Swab the fruit with alcohol and wash in sterile water. Place a drop of inoculum (10^6 cfu/ml) (5.4) on the fruit surface and puncture the fruit by pricking through the drop with a sterile needle. Incubate at 25°C in closed boxes lined with damp blotting paper. Use a known culture as a control. Inoculate fruits similarly with sterile water as a negative control.

5.3.2.1 *Erwinia amylovora*

Use immature pear fruitlets (5–8 cm long), which may be stored for several months if they are blemish-free, washed briefly in sodium hypochlorite solution, rinsed in sterile water, packed individually in new plastic bags and stored at 2–5°C. Cut transverse slices, about 1 cm thick, and place on damp filter paper in Petri dishes. Inoculate the cut surfaces with a weakly turbid suspension (approximately 10^7 cfu/ml) without delay by stabbing. Incubate for 3 days.

E. amylovora causes a rot of immature pear tissue accompanied by characteristic droplets of creamy ooze.

5.3.2.2 *Pseudomonas syringae* pv. *syringae*

Lemon fruits (preferably freshly picked) or immature pear fruitlets (see 5.3.2.1) can be used. Symptoms appear as deep, black necrotic pits on lemons and dark, watersoaked necrotic spots on pear. Many strains produce watersoaked lesions with a reddish margin on dwarf bean pods.

5.3.2.3 *Pseudomonas syringae* pv. *phaseolicola*

Characteristic watersoaked necrotic lesions are produced on immature dwarf bean pods.

5.3.2.4 *Pseudomonas syringae pv. lachrymans*

Tear drops of bacterial ooze are produced on the surface of green cucumber fruits.

5.3.2.5 *Xanthomonas campestris* pv. *juglandis*

Necrotic spots are produced on immature walnut fruits.

5.3.2.6 *Xanthomonas campestris* pv. *phaseoli*

Characteristic necrotic spots are produced on immature dwarf bean pods.

5.3.3 Host tests for canker pathogens

Stab soft stems of plants with a needle charged with culture (10^6–10^7 cfu/ml, see 5.4). Symptoms appear more reliably and quickly if the site of inoculation is protected from desiccation with polythene for 48 hours. Many pathogens will produce spreading lesions and dieback on young, detached shoots standing in water.

Lenticular cankers may be produced by incubating young shoots of plants in conditions of high relative humidity in a damp chamber or with a polythene bag tied over the stems before spraying with a suitable suspension.

The canker phase of some diseases cannot be produced by inoculation at all times of the year. For example, inoculations of cherry with *Pseudomonas syringae* pv. *morsprunorum* should be carried out in autumn and winter.

5.3.4 Host tests for leaf and stem spot pathogens

Most leaf and stem pathogens infect plants via lenticels and stomata although it is sometimes difficult to reproduce conditions for this to take place artificially. This has led to a variety of inoculation methods being used, ranging from spraying leaves with inoculum as a fine mist or under pressure through an artists' spray brush, to vacuum infiltration and inoculation after various forms of wounding.

Most of these have value for specific purposes such as breeding for resistance when their disadvantages as indicators of pathogenicity or virulence have been thoroughly studied and understood for particular host/pathogen combinations. However, some of these methods can cause reactions on non-hosts that could be misinterpreted as pathogenesis. The diagnostician is therefore often in the dilemma of choosing either a method most likely to give a pathogenic response by a compatible pathogen combination but which could give a misleading result if the plant used is not a host of the bacterium tested, or a method that avoids responses by incompatible host/pathogen combinations but often also fails to give a response with a compatible combination. Without wide experience of many host/pathogen combinations, it is preferable to adopt a compromise method that avoids much of the dilemma. This involves misting inoculum at a comparatively low concentration on to plants, a few leaves and/or stems of which are lightly wounded, and incubating at high humidity. It must be noted that some plant pathogenic bacteria are also associated with human disease. All misting should therefore be performed in a biological safety cabinet which exhausts outside the building (1.11).

Young plants should be incubated in polythene sleeves, with wires arranged to avoid the bags touching the plant, for 24 hours before inoculation (Fig. 5.2). Prepare a suitable inoculum by washing the growth from a nutrient agar or other suitable agar medium in sterile water. Mix well and dilute in sterile water to obtain a population of 10^5-10^6 cfu/ml (5.4). Prick one or two labelled leaves on each plant and the upper part of a young stem or a young petiole in about five places with a sterile needle.

Spray inoculum as a fine mist until the plant surface is wet, ensuring that upper and lower leaf surfaces are wetted and cover immediately

Fig. 5.2. Young host plant incubated in a polythene sleeve with a wire frame to prevent the plant touching the polythene.

to prevent drying out. Any low-pressure, easily sterilizable sprayer can be used; a hypodermic syringe with a needle the tip of which has been bent at right angles to the bevel by pressing it on a hard surface, is very convenient and produces a fine, easily-directed, fan-shaped spray (Fig. 5.3). If the top of the sleeve is undone the plant can be sprayed while still inside the sleeve, reducing risk of infecting other plants. Incubate for 48 hours before removing from the chamber or polythene bag.

Use plants inoculated with a virulent, known culture as a positive control and plants sprayed with water as a negative control. Grow on in a glasshouse or growth chamber at approximately 20–25°C and 85% RH with a light regime of 12–16 hours if available.

Observe for typical, watersoaked lesions developing from 3 days onwards. These should become necrotic, rounded or angular, usually brown-black and usually with a watersoaked margin. On stems, lesions are often elongated along the stem. If lenticular or stomatal infections do not occur,

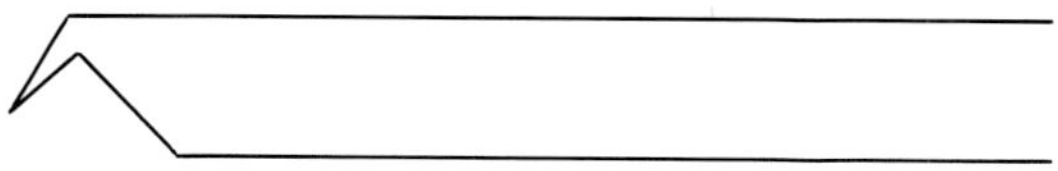

Fig. 5.3. Hypodermic needle with its tip bent at right angles to produce a baffle for mist production.

wound infection usually succeeds in producing symptoms but caution is then needed in interpreting the result. Repeat tests using different methods and conditions may be necessary to ensure accurate diagnosis. The use of a positive control is essential for determining that the conditions of inoculation have the potential for producing infection and symptom expression.

This method can be used successfully for many leaf spot and stem spot pathogens, including:

Curtobacterium flaccumfaciens pv. *oortii* (see 5.3.4.5)	Tulip
C. flaccumfaciens pv. *betae* (see 5.3.4.5)	Beet
Erwinia amylovora	Apple, pear, cotoneaster
Pseudomonas syringae pv. *lachrymans*	Cucumber
P. s. pv. *phaseolicola*	Bean
P. s. pv. *syringae*	Lilac, pear
P. s. pv. *morsprunorum*	Cherry, plum
P. s. pv. *tabaci*	Tobacco
P. s. pv. *glycinea*	Soybean
P. s. pv. *pisi*	Pea
Pseudomonas andropogonis (syn. *P. woodsii*)	Carnation
P. avenae	Grasses
P. andropogonis	Grasses
P. marginalis	General
P. viridiflava	General
Xanthomonas fragariae	Strawberry
X. campestris pv. *campestris*	Brassicas
X. c. pv. *citri*	Citrus
X. c. pv. *pruni*	Plum
X. c. pv. *oryzae*	Rice
X. c. pv. *oryzicola*	Rice
X. c. pv. *manihotis*	Cassava
X. c. pv. *malvacearum*	Cotton

See also 5.3.4.1–5

5.3.4.1 *Pseudomonas corrugata*

Inject inoculum (10^8 cfu/ml) into the pith of 9–12 inch high, pre-wilted plants using a hypodermic syringe. Cover with a polythene bag and incubate for 48 hours before growing on in a glasshouse.

Symptoms include black, necrotic lesions on stems and necrosis of the pith up the stem usually with a characteristic browning of the stele and laddering of the pith.

5.3.4.2 *Clavibacter michiganense* subsp. *michiganense*

This test uses young, fully-expanded cotyledons, although young expanding leaves may be used. The symptoms formed are similar.

Sow tomato seeds of a susceptible cultivar (e.g. cv. *Moneymaker*) 10–12 seeds to a pot. Thin out seedlings to five, well-spaced in the pot. Prepare an aqueous bacterial suspension (approximately 10^7 cfu/ml). Add a drop to a cotyledon and abrade the epidermis by rubbing twice firmly with the forefinger. The epidermis may be abraded prior to introducing the inoculum which may be spread over the surface with a sterile cotton wool swab. Carborundum may be used to assist in the epidermal abrasion but tends to cause a general necrosis and is not necessary. Repeat for other cotyledons and cover the pot with a polythene bag. Incubate for 48 hours and remove the bag. *C. michiganense* subsp. *michiganense* causes white blister-like craters which usually develop within 4 days. As they expand the epidermal surface may become ruptured, especially along the midrib.

5.3.4.3 Other tomato pathogens

Cotyledon tests (5.3.4.2) may also be used for other tomato diseases. Characteristic spots and scabs are formed with *Xanthomonas campestris* pv. *vesicatoria* and pin-prick-sized speck lesions are formed with *Pseudomonas syringae* pv. *tomato.*

5.3.4.4 *Pseudomonas syringae* pv. *glycinea*

This test uses fully expanded soybean cotyledons. Place a droplet of bacteria (10^6 cfu/ml, see 5.4) near the margin of the cotyledon and puncture it through the droplet. Incubate inoculated plants at *c.*25°C, covered with a polythene bag to maintain high humidity. A positive cotyledon test is given by the development of a dark-brown necrosis around the wound. The underside of the spot may produce a pale-grey slime which may become papery when dry.

5.3.4.5 **Leaf abrasion for *Curtobacterium flaccumfaciens* pvs. *betae* and *oortii***

Inoculate leaves of the appropriate host (*Tulipa* spp. for, *C. flaccumfaciens* pv. *oortii*, red beet cv. *Cheltenham Greentop* for *C. flaccumfaciens* pv. *betae*) by rubbing gently with carborundum powder before spraying on an aqueous suspension of the test bacteria (10^6-10^7 cfu/ml, see 5.4). Cover with a polythene bag. Observe for the development of silvery, roughened patches containing masses of Gram-positive coryneform bacteria.

5.3.5 Host tests for vascular pathogens

Bacteria should be introduced directly into the vascular tissue and this can be done in different ways depending on the host/pathogen combination.

5.3.5.1 **Stem inoculation**

Place a droplet of a bacterial suspension (approximately 10^6-10^7 cfu/ml, see 5.4) on to the stem surface, usually in a leaf axil if convenient. Stab through the droplet into the vascular tissue with a sterile needle taking care not to pierce the opposite side of the stem. Alternatively, the eye of a large sewing needle which is charged with bacteria may be stabbed through the stem in to the vascular tissue. Cover the plant with a polythene bag for 24 hours to prevent the droplet or the wound from drying out. Incubate at temperature appropriate to the disease and observe the plants for up to several weeks or even months for wilting, discoloration, blocking and necrosis of vascular tissue and/or ooze production.

5.3.5.2 **Inoculation of cuttings**

Stand cuttings in a bacterial suspension (10^6-10^7 cfu/ml, see 5.4). for a few minutes before striking in a suitable sterilized potting medium and incubating for symptom production (5.3.5.1).

5.3.5.3 **Root inoculation**

Lift young plants from pots and wash the root systems clean of soil. Trim the roots with scissors or a knife and dip into a suitable bacterial suspension (10^6-10^7 cfu/ml) for 5–10 minutes. Plants may be placed in a strong air current during this period to facilitate transpiration and uptake of bacterial cells. Plants should be potted in a suitable sterile growth medium.

5.3.5.4 Inoculation of cut cotyledons of lucerne seedlings for *Clavibacter michiganense* subsp. *insidiosum*

Grow seedlings of a susceptible lucerne cultivar (e.g. Du Puits or Grimm) in seed boxes or pots. Dip scissors into a bacterial suspension (10^6-10^7 cfu/ml, see 5.4) and cut the cotyledonary leaves. Cover with a polythene bag for 24 hours, remove and grow on. Observe for wilting.

5.3.5.5 Eggplant challenge test for *Clavibacter michiganense* subsp. *sepedonicum*

Where latent infections of *Clavibacter michiganense* subsp. *sepedonicum* are suspected in potato tuber lots, the eggplant challenge test provides the most accurate test for confirmatory diagnosis (3.5.7)

Eggplants should be not more than 100 mm high and at the third leaf stage. Twenty-five plants should be inoculated with the pellet (3.5.7) obtained from each tuber sample (e.g. 200 tubers). Since it is difficult to inject stems due to the high turgor pressure, they should be slit with a scalpel just above the cotyledons. The slit should be vertical or slightly diagonal, approximately 1 cm long and approximately deep enough to penetrate the vascular tissue (about 1 mm) but not enough to pierce the opposite side of the stem. Turn the scalpel blade to open the slit and place the pelletted inoculum in it with a sterile, fine paint brush or eyeliner brush. Seal the wound with sterile Vaspar or petroleum jelly which can be conveniently applied from a 2 ml plastic hypodermic syringe (without a needle).

Incubate the plants at approximately 21°C ensuring that the temperature does not rise above 30°C. Symptoms can appear within 6 days but at very low concentrations of inoculum they may take over 3 weeks to develop, so plants should be incubated 40 days before a negative result is recorded.

Symptoms include a partial leaf wilt usually restricted to one sector of leaf which usually originates at the leaf-edge. Infection may also be seen as greasy spots within the leaf lamina (Fig. 5.4). As the infected leaves age the margins of the flaccid areas may become chlorotic. Plants incubated under favourable growth conditions may not die. With experience the typical, sectored, leaf wilts are readily distinguished from the more general wilts caused by *Phoma exigua*, *Erwinia carotovora* or simply by large populations of saprophytic bacteria. With these organisms the whole leaf and petiole usually wilt. Before diagnosing *Clavibacter michiganense* subsp. *sepedonicum* the presence of large numbers of Gram-positive coryneform cells should be demonstrated in the wilted leaf tissue and in the main stem just above the point of inoculation.

Fig. 5.4. Sectored wilting of an eggplant leaf by *Clavibacter michiganense* subsp. *sepedonicum*. Greasy spots within the lamina can also be seen. Crown copyright. Used by permission of the Controller of HMSO.

The eggplant challenge test may be used to confirm the identity of *C. michiganense* subsp. *sepedonicum* cultures. Use the same method of inoculation with an inoculum of c.10^8-10^9 cfu/ml on at least ten eggplants. Symptoms usually develop within 6–20 days.

5.3.6 Host tests for soft-rotting bacteria

Storage organs, e.g. bulbs/tubers or succulent stems provide the best host tissue for soft rot tests. When possible the original host should be used. Tissue slices are often used for host tests although regard should be paid to the fact that some pectolytic bacteria which may rot slices may only rot whole plant organs under unusual conditions. Some opportunist pathogens such as *Bacillus polymyxa* and *Cytophaga* sp. will rot slices.

Potato tubers, cucumber fruits and onions are useful organs for soft rot tests. If used whole, they should be washed and disinfected by wiping with alcohol or dipping in 0.5% sodium hypochlorite solution. Make a wound either by inserting a sterile cork borer to a depth of 0.5–1.0 cm into the tissue or by removing a small piece of tissue. A suitable bacterial suspension (approximately 10^7 cfu/ml) should be applied to the cut surface which should be sealed with Vaspar or paraffin wax. Tubers should be incubated in the dark. A humid environment will promote rotting.

Tissue slices must be cut aseptically after carefully peeling, and it must be remembered that potato tuber lenticels are often infected with soft rot bacteria so several uninoculated controls are essential when using potato tubers or slices.

Place the slice on moistened, sterile filter paper in a Petri dish or similar receptacle. Incubate at 25°C after inoculating it with a suitable suspension (10^7 cfu/ml). This is best added along a groove or nick cut in the slice. Test for rotting by probing with a sterile needle. Rotting by *Pseudomonas* spp. is usually slower than by *Erwinia* spp.

Another useful method for inoculating whole organs with soft rot bacteria is by pricking the organ surface with a sterile toothpick dipped into a dense bacterial suspension. Incubate tubers in a dark, humid chamber.

5.3.7 Host tests for gall producing bacteria

5.3.7.1 ***Pseudomonas syringae* subsp. *savastanoi***

Stab young leaf, petiole or stem tissue through a droplet of bacterial culture (approximately 10^7 cfu/ml, see 5.4). Prevent desiccation either by covering the wound with Vaspar or petroleum jelly or by covering the tissue with a polythene bag.

Isolates from ash should be introduced into young ash stems, isolates from olive into young olive stems and isolates from oleander into young oleander leaves, petioles or stems.

5.3.7.2 ***Agrobacterium tumefaciens***

Agrobacterium tumefaciens isolates usually have a wide host range but no host is affected by more than about 80% of strains (Anderson & Moore, 1979). Use the original host where this is possible. If this is impractical, several hosts should be selected. The following are useful and easily-cultivated hosts: tomato, marigold, sunflower, bryophyllum, chrysanthemum.

Inoculate either by stabbing young stems or leaves with a needle or toothpick charged with culture from a plate or by stabbing through a droplet of bacterial suspension (approximately 10^8 cfu/ml, see 5.4). Galls usually develop within 2 weeks.

Isolates from galls on rosaceous hosts may not readily produce symptoms.

Root slice test for Agrobacterium spp. This test is useful for many but not all pathogenic strains of agrobacteria. Use carrot, turnip, beetroot or parsnip. The roots should be freshly lifted, young and undamaged. Wash by scrubbing with a soft nail brush. Immerse in a strong solution of sodium hypochlorite (10 000 ppm available chlorine, or 1:10 dilution of a commercial sodium hypochlorite will 10% available chlorine) for 10 minutes. Wash under running tap water and blot dry. Dip the roots in 70% ethanol and flame twice.

Place the roots on a sterilized tile and slice with a sterile knife into discs about 5 mm thick, discarding the root and crown ends. Place the slices in Petri dishes containing sterile filter paper. Cut the slices into halves and add 5 ml of sterile water to wet the paper.

Place a few drops of sterile water on each half of each root slice. Spread the liquid over the entire surface of one half; this acts as a control. A large loopful of growth from a 48-hour culture grown on nutrient dextrose agar should be rubbed over the surface of the other half slice. Use a known culture as a positive control. Wrap the Petri dish in cling-film and incubate at room temperature for up to 30 days.

The development from the cambium of small galls is diagnostic for tumorigenic and rootlets for rhizogenic agrobacteria. Symptoms usually appear in 12–15 days. Record a negative result if no symptoms appear on slices inoculated with the test bacterium but symptoms appear on those inoculated with the known culture. If no symptoms appear with either record an indeterminate result.

Red beetroot is particularly useful for *A. rhizogenes*. For further details see Ark and Thompson (1961).

5.3.7.3 Leafy gall produced by *Rhodococcus fascians*

Rhodococcus fascians can cause fasciation on a wide variety of hosts, but isolates often have quite narrow host ranges. Sweet pea or black gram (*Cicer arietinum*) seedlings are useful hosts, but where possible the original host should be used.

Young tissue may be stabbed as for agrobacteria (5.3.7.2) or cuttings from suitable hosts may be dipped in a bacterial suspension (approximately 10^7 cfu/ml) (5.4). Symptoms may take over a month to develop.

However, the following laboratory test is rapid and simple and gives unambiguous results with all but a few *R. fascians* strains.

Sweet pea test. Shake 12 seeds in 0.1% Manoxol to remove adherent air bubbles. Decant the wetter and surface disinfect by immersing the seeds in a 1:100 dilution of a household bleach or 2% Chloramine T for 10 minutes. Decant and wash three times with sterile, distilled water. Transfer the seeds aseptically to sterile, damp blotting paper in a sterile Petri dish. Incubate at 25°C until the plumules are about 1 cm long (5–7 days). Dip each seedling in a heavy suspension of bacteria in sterile, distilled water (*c.*10^8 cfu/ml) (5.4) and aseptically transfer to test tubes containing 2–3 cm of moist, sterile sand. Add sufficient bacterial suspension to just cover the sand. Inoculate six seeds with the unknown culture, four with a known culture and leave two as negative controls. Incubate at room temperature in the light. The production of fasciation from the base of the seedlings is a positive result. Only record a negative result if seedlings similarly inoculated with a known virulent culture of *R. fascians* produce symptoms. Most, but not all, strains of the pathogen are pathogenic to sweet pea.

5.3.8 Host tests for bacteria causing mushroom diseases

The best method for testing the potentiality of a bacterium for inciting a mushroom disease is to impregnate the casing with the suspect bacterium at a concentration of $>10^7$ cfu/ml (see 5.4) before applying the casing to the bed. Adequate controls, e.g. a known culture and a sterile water control are essential. However, such tests are usually difficult or impractical for the diagnostician to perform and detached cap tests must often be relied upon. Tests on detached caps are not really very suitable since the resistance or susceptibility of a detached cap is likely to be different from that of an actively-growing cap. If these tests are used it is essential to use freshly-picked caps or preferably actively-growing sporophores. Prepare a 10^7–10^8 cfu/ml suspension (see 5.4) from a nutrient agar culture. Add drops of the suspension to some caps and paint a cross with the suspension on others. Use sterile water controls and where possible, known culture controls. For bacterial pit tests prick some of the caps through the droplets.

Incubate in a humid chamber for 2 days and observe for browning of the cap in the inoculated areas (not found with the water controls) (*Pseudomonas tolaasii*, *P. gingeri*) and depressions or pits (*Bacillus polymyxa*). Where drippy gill caused by *Pseudomonas agarici* is suspected inject the undersurface of unopened caps or very young gills and incubate as above. Observe for droplets of ooze and watersoaking of the gills. Wong & Preece (1979) developed a rapid test using mushroom blocks which they believe is an improvement upon conventional cap tests.

5.4 Inoculum

Ideally inoculum for host pathogenicity tests should contain a known number of highly virulent cells. However, the achievement of this ideal usually requires considerable research for each pathogen and perhaps for each host/pathogen combination. The diagnostician is forced to adopt a more pragmatic approach and to try to avoid the more obvious pitfalls, but still to do host tests as rapidly as possible.

To obtain suspensions for inoculation of aerobic pathogens of fairly uniform physiologic state and free of culture metabolites that might interfere with or distort host reaction, it is usually necessary to grow cultures in liquid, shake-culture and then wash several times in water. This can, however, render cultures susceptible to shock and death, which can be extremely variable and uncontrollable, if diluted in sterile, distilled water. If other diluents are used instead — buffers or low concentrations of protein — the problem can usually be overcome, but at the expense of introducing unwanted and artificial factors into the inoculum. One satisfactory way of overcoming dilution shock with *Erwinia amylovora* is to dilute in a low concentration (10^4/ ml) of heat-killed (60°C for 10 minutes) cells of the same bacterium which probably allows inoculum to be produced that simulates natural inoculum as it is dispersed by rain splash. This method is too time consuming for most routine diagnostic laboratories and it is usually sufficient to prepare inoculum by washing visible growth from young agar culture with sterile, distilled water and then diluting in sterile, distilled water. Dilution shock is not known with unwashed cells grown on agar and, if the suspension is removed quickly from the agar, little of the metabolic products are likely to be removed from it with the culture growth.

To prepare inoculum of given concentrations of viable cells it is necessary to relate optical density (od) of suspensions to numbers of viable cells or, more accurately, to numbers of colony-forming units (cfu) per unit volume. The latter takes some days to determine, by which time the concentration of cfu's in the suspension is likely to have changed, so it is necessary to prepare a plot of od against cfu/ml from which cfu/ml of a suspension can be estimated immediately and the suspension then adjusted to the required concentration for inoculation. Since the slope of this plot may vary between pathogens (and even between different isolates of the same pathogen) and with growth conditions of the culture, it is necessary for research purposes to construct a plot for each isolate and for differing growth conditions. However, for diagnostic purposes, the cfu concentration is not so critical as to require more than a representative plot for a group of closely-related pathogens if the growth conditions are standardized. For preparing such plots proceed as follows:

Grow a freshly-isolated culture on appropriate agar medium until growth is visible. Wash off in 5 ml sterile, distilled water, mix well on a mechanical mixer and prepare two-fold dilutions in sterile, distilled water until the suspension is just turbid. Determine the optical density of each dilution on a spectrophotometer at 600 nm. Plot od against dilution on log/log paper. From the straight part of the plot choose two dilutions and from each of these prepare a ten-fold dilution series from 10^{-1} to 10^{-7}. From each of the dilutions 10^{-4}, 10^{-5}, 10^{-6} and 10^{-7} pipette 0.2 ml on to each of 5 plates of the same medium and spread with a bent glass rod (preferably on a turntable). The plates should be dried sufficiently so that after spreading they dry again within 5–10 minutes.

Incubate and count the number of colonies on plates of the dilution that has between 40 and about 400 colonies/plate; using sufficient plates to count about 400 colonies will give acceptable standard errors. Calculate the mean equivalent number of cfu/ml for each of the dilutions for which od was recorded and then plot log od against log cfu/ml. By determining the od for a suitable suspension of any culture of these groups grown in the same way an approximate value for its cfu/ml can be obtained.

Plots should be prepared in similar ways against each of the following groups: group I and II fluorescent pseudomonads using 24–36-hour nutrient agar cultures; group III, IV and V fluorescent pseudomonads using 24–36-hour nutrient agar cultures; non-fluorescent pseudomonads using 24–36-hour cultures on sucrose peptone agar; *Xanthomonas campestris* using 24–36-hour nutrient dextrose agar cultures; *Curtobacterium flaccumfaciens* pvs. *flaccumfaciens*, *betae*, *oortii* and *poinsettiae* using 24–36-hour cultures on nutrient dextrose agar; *R. fascians* using 24–36-hour cultures on nutrient dextrose agar; *Clavibacter rathayi*, *C. iranicum*, *C. michiganense* subsps. *michiganense*, *sepedonicum* and *insidiosum* using 40–60-hour cultures on nutrient dextrose agar; *Erwinia carotovora*, *E. chrysanthemi* using 24–36-hour cultures on nutrient agar; *E. amylovora* using 24–36-hour cultures on nutrient agar. Other pathogens cannot be conveniently grouped and the relation of od for cfu/ml for these may need to be determined separately. For critical work inoculum should be stood in ice if many inoculations are being made from one suspension to avoid any fall in concentration of cfu/ml during the period. For research work it is often necessary to determine the cfu/ml of the inoculum at the beginning and end of a series of inoculations.

6

Media and Methods

6.1 Cell morphology

6.1.1 General purpose staining

The most useful stain for differentiating and visualizing bacteria in the presence of plant cell debris is methylene blue. It is particularly useful if the facility for phase-contrast microscopy is not available and in some cases gives better differentiation than the latter, especially if the proportion of debris to bacterial cells is high. Loeffler's form of the stain gives consistently good results.

Loeffler's methylene blue staining

1 Solution A.
- **a** Methylene blue (90% dye content) 0.3 g
- **b** Ethyl alcohol (95%) 30.0 ml

2 Solution B.
- **a** KOH (1% aqueous solution) 1.0 ml
- **b** Distilled water 99.0 ml

Mix solutions A and B and store in a bottle not more than half-full.

The stain ripens by oxidation, which takes several months to complete, and improves with further keeping. Plug the bottle with a loose cotton wool plug and shake it frequently while it is ripening. Prepare enough to last 5–10 years.

Method of use. Make faintly turbid suspension in sterile water of bacterial growth from a solid medium. Smear a loopful of this suspension or of suspended plant material on a clean, grease-free slide. Allow to dry and then fix the smear by passing the slide rapidly two or three times through a Bunsen flame. Cover the smear with stain for 1 minute, pour off the stain and wash in a gentle stream of water until no more stain is obviously being removed from the smear (usually about 5–10 seconds). Dry by blotting on filter paper and examine under oil immersion. Cells stain varying shades of blue. Beading, barring and granules in cells of coryneform bacteria stain more deeply and spores appear as unstained bodies within blue cells.

6.1.2 Gram reaction

One of the major divisions in the eubacteria (the group to which most bacterial plant pathogens belong) is that between those that can retain a complex of crystal violet and iodine against elution with ethanol (Gram-positive) and those that cannot (Gram-negative). This reflects fundamental biochemical and biophysical differences in the bacterial cell wall. The reaction is useful as a primary determinative character for many plant bacteria.

6.1.3 Gram stain (Hucker modification)

Hucker's crystal violet

Dissolve 2 g crystal violet in 20 ml 95% ethanol. Dissolve 0.8 g ammonium oxalate in 80 ml distilled water. Mix the two solutions.

Lugol's iodine (Gram's modification)

Iodine	1.0 g
KI	2.0 g
Distilled water	300 ml

Grind together the solids in a pestle and mortar and then dissolve in the water and stir in a closed container for several hours to complete dissolution.

Safranin counterstain

Prepare a stock solution by dissolving 2.5 g safranin 0 in 100 ml 95% ethanol. Dilute this 1:10 in distilled water for use.

Method of use. Make a fairly turbid suspension of bacterial growth from a solid medium in sterile water. Smear a loopful of this suspension or of suspended plant material on a slide. Allow to dry then fix by passing the slide rapidly two or three times through a Bunsen flame. Flood the slide with crystal violet for 1 minute. Wash in a gentle stream of tap water until no more stain is obviously being removed from the smear and flood with Lugol's iodine for 1 minute. Wash in a gentle stream of tap water and blot dry. Decolorize by washing in a gentle stream of 95% ethanol for not more than 30 seconds to remove any stain that will easily wash away, and blot dry. Counterstain by flooding with safranin for 20 seconds. Wash in tap water, blot and examine.

Interpretation of results. Gram-positive bacteria stain blue-violet and Gram-negative bacteria stain pink-red. A few otherwise Gram-positive bacteria (mainly Bacillus spp.) sometimes give a variable or a Gram-negative reaction to staining. Where this is suspected, the Gram reaction may be determined by a potassium hydroxide solubility test. Homogenize a loopful of a well-grown colony in a drop of 3% aqueous KOH on a glass slide. Raise the loop a few centimetres. If strands of viscid material are seen the bacterium is Gram-negative. If such strands are not seen on repeated strokes of the loop the bacterium can be considered to be Gram-positive.

6.1.4 Flagella

Most plant pathogenic bacteria are motile; exceptions include certain coryneform species and *Erwinia stewartii*. All motile species use flagella except for pectolytic *Cytophaga* spp. which are gliding bacteria. Motility may be observed in normal wet mounts or hanging drops by phase contrast microscopy but if present is often so only in early log phase of the growth of cultures. This is usually after about 8–18 hours incubation. Flagella can be observed by light microscopy or electron microscopy. Electron microscopy is a more reliable and less time-consuming method. Determination of presence and arrangement of flagella is not needed for routine diagnosis of plant bacteria but methods to do this are available in general texts such as Mackie & McCartney, (1978).

6.1.5 Poly-β-hydroxybutyrate accumulation

Poly-β-hydroxybutyrate (PHB) granules are accumulated by many species of non-fluorescent pseudomonads, especially in media with high carbon: nitrogen ratios. This accumulation is a useful diagnostic test. There are several methods for its detection, although because most tests do not identify the polymer specifically it is important to use known positive and negative control cultures when using these methods.

Methods for detection. Grow cultures on 5% sucrose nutrient agar or sucrose peptone agar or preferably in shake-culture in a basal medium (Ayers, Rupp & Johnson's Base (6.3.2) containing 0.5% β-hydroxybutyrate (sodium salt)). Production of PHB is more certain in the latter medium. When good growth has been obtained (after about 2 days) observe smears or wet preparations by phase contrast microscopy and look for granules which are hyaline, circular, one or more to a cell and occupying much of the cell volume. PHB granules also show up well in the electron microscope as electron-dense bodies. Cells which have accumulated PHB granules do not usually Gram stain well. Since spores cannot be easily differentiated from polymer granules by this method, it should only be used when it is known from other evidence that the culture being observed is of a non-spore-forming bacterial species.

Alternatively, flood a heat-fixed smear (6.1.1) with Nile blue A (1% aqueous solution) for 10 minutes at 55°C. Wash briefly with tap water. Flood with 8% aqueous acetic acid for 1 minute to remove excess stain. Wash again in tap water and blot dry (for more detailed methods of preparing smears see 6.1.1). Remoisten with a drop of water and cover with a coverslip. Examine under oil immersion with epifluorescence at 450 nm. PHB granules fluoresce bright orange (Ostle & Holt, 1982).

If only light microscopy is available granules can be stained with Sudan black B, although this stain does not always differentiate PHB granules satisfactorily. Prepare a 0.3% solution of Sudan black B in 70% ethanol. Allow to stand for 12 hours after the dye has dissolved. Flood a heat-fixed smear (6.1.1) with the dye for 10 minutes. Drain, blot dry and flood briefly with xylol or water to clear the slide. Blot dry and counter-stain with 0.5% aqueous safranin for 10 seconds. Wash off in tap water, blot dry and examine under oil immersion. PHB granules stain blue-black.

6.1.6 Endo-spore formation

Smear a loopful of bacterial suspension on a clean microscope slide. Allow to dry and thoroughly heat fix (6.1.1). Stain for 10 minutes in a saturated solution of Malachite green. Wash in gently-running tap water. Counter stain with 0.25% safranin (see 6.1.3) for 15 seconds. Wash with water. Blot dry and examine under the oil immersion objective. Spores appear bright-green within pale-pink cells.

6.2 Isolation and growth media

6.2.1 Introduction

Relatively few selective media are of great use in the isolation of bacterial plant pathogens except from old or badly contaminated specimens and from soil, and only those whose value in isolation and diagnosis is well-proven are given here. Non-selective media which allow good growth with characteristic colony morphology are generally recommended (see Chapter 3). Few plant pathogenic bacteria are fastidious and good growth can usually be obtained on relatively simple media. Generally speaking, although many plant bacteria grow better on media containing added utilizable carbohydrate such media are not suitable for maintaining them unless chalk is added, since acids often accumulate in the media as by-products of utilization to an extent that they become bactericidal. With few exceptions, nutrient agar or yeast dextrose chalk agar are preferred maintenance media.

6.2.2 Brisbane and Kerr's media for *Agrobacterium* biovars (Brisbane & Kerr, 1983)

Medium 1A for A. tumefacians biovar I

L(—) arabitol	3.04 g
NH_4NO_3	0.16 g
KH_2PO_4	0.54 g
K_2HPO_4	1.04 g
$MgSO_4.7H_2O$	0.25 g
sodium taurocholate	0.29 g
crystal violet (0.1% aqueous solution)	2.0 ml
Davis agar	15.0 g
distilled water	1 litre

Mix, heat to dissolve, dispense into 100 ml amounts and sterilize by autoclaving at 121°C for 15 minutes. Before pouring add per 100 ml:

2% aqueous actidione	1 ml
1% aqueous sodium selenite	1 ml

Medium 2E for biovar II

erythritol	3.05 g
NH_4NO_3	0.16 g
KH_2PO_4	0.54 g
K_2HPO_4	1.04 g
$MgSO_4.7H_2O$	0.25 g
sodium taurocholate	0.29 g
yeast extract (1% aqueous solution)	1.0 ml
Malachite green	5.0 ml
Davis agar	15 g

Mix, heat to dissolve, dispense into 100 ml amounts and sterilize by autoclaving at 121°C for 15 minutes.

Before pouring add per 100 ml:

2% aqueous actidione	1 ml
1% aqueous sodium selenite	1 ml

Medium 3DG for biovar III

Solution A	
(+)−sodium tartrate	5.75 g
4% aqueous D-glutamic acid (neutralized to pH 7 with NaOH)	15.0 ml
$NaH_2PO_4.2H_2O$	6.24 g
$Na_2H\ PO_4$	4.26 g
NaCl	5.84 g
$MgSO_4.7H_2O$	0.25 g
sodium taurocholate	0.29 g
1% aqueous yeast extract	1.0 ml
0.1% aqueous congo red	2.5 ml
distilled water	500 ml
Solution B	
$MnSO_4.4H_2O$	1.12 g
Davis agar	15.0 g
distilled water	500 ml

Mix, heat to dissolve, dispense solutions A and B separately in 50 ml amounts and sterilize by autoclaving at 121°C for 15 minutes.

Before pouring add per 50 ml of solution B:

2% aqueous actidione	1.0 ml
1% aqueous sodium selenite	1.0 ml

Add solution A at 50°C. Redistribute the precipitate that forms by shaking gently.

Agrobacteria grow well on these media as white, domed, mucoid colonies. Other bacteria may grow but colonies are small.

6.2.3 Clark's medium (Clark, 1969) — a semi-selective medium for *Agrobacterium* spp.

Base

lactose	5.0 g
KNO_3	1.0 g

$MgSO_4.7H_2O$	0.1 g
Na_2HPO_4 (anhydrous)	0.18 g
Oxoid agar No. 3	12.0 g
distilled water	1 litre

Adjust to pH 6.8. Sterilize by autoclaving at 121°C for 20 minutes. Prepare the following aqueous solutions:

0.25% (w/v) ferric sodium sequestrate
33.5% (w/v) $MnSO_4.4H_2O$

Add 1 ml of each solution to 1 litre of molten base (45°C), mix and pour plates.

Agrobacteria appear as pearly-white, domed colonies within 3 days.

6.2.4 Doepel's medium

Clavibacter rathayi is a fairly fastidious organism and although it will grow on nutrient dextrose agar, growth is often sparse. Doepel's medium gives better growth.

Doepel's medium

glucose	10.0 g
casein hydrolysate	8.0 g
yeast extract	8.0 g
K_2HPO_4	2.0 g
$MgSO_4$	0.3 g
agar	12.0 g
distilled water	1 litre

Sterilize by autoclaving at 121°C for 15 minutes.

6.2.5 Glycerol agar

A suitable medium for producing motile cultures for flagella staining or for flagella antigens, or for 'phage tests.

glycerol	20 ml
yeast extract	3.0 g
peptone	5.0 g
oxoid agar No. 3	12.0 g
distilled water	1 litre

Adjust to pH 7.0 if necessary. Sterilize by autoclaving at 121°C for 15 minutes.

6.2.6 Kerr's medium (New and Kerr, 1971)

A commonly used medium for the isolation of *Agrobacterium* biovar II. This medium is not wholly selective for this biotype but is a useful semi-selective medium.

Base

erythritol	5.0 g
$NaNO_3$	2.5 g
KH_2PO_4	0.1 g
$CaCl_2$	0.2 g
NaCl	0.2 g
$MgSO_4.7H_2O$	0.2 g
0.65% aqueous ferric sodium sequestrate	2.0 ml
biotin	2.0 μg
Oxoid agar No. 3	12.0 g
distilled water	1 litre

Adjust pH to 7.0, dispense in 95 ml volumes and sterilize by autoclaving at 121°C for 15 minutes. Add the following filter-sterilized antibiotic solutions to the molten medium at 50°C and mix well.

sodium selenite (0.1 g/ml)	1.0 ml
actidione (0.01 g/ml)	2.25 ml
bacitracin (650 units/ml)	1.0 ml
tyrothricin (0.001 g/ml)	0.1 ml

Agrobacterium biovar II colonies are domed, entire and pearly-white often becoming tan-coloured after 4–5 days.

6.2.7 King's medium B

Although King's medium B (King *et al.*, 1954) is commonly used for fluorescein pigment production, not all fluorescent pseudomonads readily produce pigment on the medium, but they may do so on caesamino-sucrose-gelatin agar (6.2.8) or Oxoid Pseudomonas F medium.

King's medium B

proteose peptone (Difco No. 3/Oxoid L46)	20.0 g
glycerol	10.0 g
K_2HPO_4	1.5 g

$MgSO_4.7H_2O$	1.5 g
agar	15.0 g
distilled water	1 litre

Adjust pH to 7.2 and sterilize by autoclaving at 121°C for 15 minutes. (Do not filter the medium).

Green-fluorescent pseudomonads produce a diffusible yellow, green or brownish pigment, which fluoresces green or blue in ultraviolet light.

6.2.8 Caesamino-sucrose-gelatin medium

This medium has been used for green-fluorescent pigment production of *Pseudomonas syringae* pv. *morsprunorum* and *P. syringae* pv. *persicae* instead of King's medium B (6.2.7).

vitamin-free caesamino acids (Difco)	10.0 g
sucrose	10.0 g
$MgSO_4.7H_2O$	1.0 g
gelatin	30.0 g
agar	20.0 g
distilled water	1 litre

Adjust to pH 7.0. Sterilize by autoclaving at 121°C for 20 minutes.

6.2.9 Nutrient agar

Commercially available preparations such as Oxoid (CM3) or Difco are usually preferable to nutrient agars prepared in the laboratory from raw ingredients.

6.2.10 Nutrient dextrose agar/broth

And 1% (w/v) D-glucose to Oxoid nutrient agar (CM3) (or Oxoid nutrient broth (CM1) to prepare nutrient dextrose broth). Sterilize by autoclaving at 121°C for 15 minutes.

6.2.11 Nutrient yeast dextrose agar

This medium is suitable for the growth of reasonably fastidious bacteria such as *Clavibacter michiganense* subsp. *sepedonicum* and *Erwinia tracheiphila*.

peptone	10.0 g
Lab-lemco	5.0 g
glucose	10.0 g

yeast extract	1.0 g
Oxoid agar No. 3	12.0 g
distilled water	1 litre

Adjust pH to 7.4 approximately. Sterilize by autoclaving at 121°C for 15 minutes.

6.2.12 Pectate gel media

A variety of pectate media are available for isolation of pectolytic bacteria. Two are given here. Paton's pectate medium is a general purpose medium. Cuppel and Kelman's CVP medium is selective for Gram-negative bacteria. On both media pectolytic *Erwinia* spp. produce deep pits with liquefaction, pectolytic pseudomonads produce shallow pits, usually without liquefaction. Paton's medium has the advantage that plates of the base medium can be poured and stored for some weeks; the overlayer can then be added the day before use. Cuppell & Kelman's CVP medium must be freshly prepared but has the advantage of being semi-selective. It should be preferred where rots are advanced and likely to contain many secondary bacteria.

6.2.12.1 **Paton's medium (modified)** (Paton, 1958)

Base

peptone	5.0 g
Lab-lemco	5.0 g
calcium lactate	5.0 g
Oxoid agar No. 3	12.0 g
distilled water	1 litre

Adjust pH to 7.2 and sterilize by autoclaving at 121°C for 15 minutes. Pour plates.

Pectate Overlayer

sodium polypectate*	2.0 g
distilled water	100 ml
disodium ethylenediamine tetra acetate (EDTA Na_2 salt)	0.01 g

*Obtainable from Ral Tech Scientific Services, Inc., 3301 Kinsman Boulevard, P.O. Box 7545, Madison, Wisconsin 53707, USA.

bromothymol blue	trace to give colour

Heat the water to 80°C on a hot plate and add the polypectate, a little at a time, with constant stirring. Add the EDTA Na_2 and maintain at 80°C. Stir constantly until the polypectate has dissolved. Add sufficient bromothymol blue to just colour the medium. Adjust the pH to 7.2 and dispense in 5 ml volumes into bijou bottles. Sterilize by autoclaving at 115°C for 4 minutes.

When the plates have set, pour on 5 ml of the cold overlayer. Allow to spread and leave for 4 hours to set. Dry (3/4 days at 25°C, or 12 hours at 37°C) before use.

The medium is fragile and should be spread with a sterile bent glass rod rather than a loop.

Pectolytic pseudomonads often produce extremely shallow pits which cannot be detected after more than several days of incubation. Plates should be looked at by oblique light and at one or two days after inoculation. If in doubt, transfer to a potato slice (5.3.7).

6.2.12.2 **CVP medium** (Cuppels & Kelman, 1974)

Preheat a blender. Add 500 ml boiling distilled water. Add the following at low blending speed:

0.07% (w/v) aqueous crystal violet solution	1.0 ml
1 mol NaOH	4.5 ml
10% (w/v) aqueous $CaCl_2.2H_2O$ (freshly prepared)	3.0 ml
Difco Agar	2.0 g
$NaNO_3$	1.0 g

Blend at high speed for 15 seconds, then continue blending slowly and add 9 g sodium polypectate (see 6.2.12.1 for source) and blend at high speed for a further 15 seconds. Pour into a 2 litre flask, cap with foil and sterilize by autoclaving at 121°C for 15 minutes. Pour plates and allow to dry well before use.

6.2.13 Potato dextrose agar

A useful medium for characterization of *Erwinia chrysanthemi* strains which produce 'fried egg colonies'.

glucose	20.0 g
peeled washed potato	200.0 g

Oxoid agar No. 3	15.0 g
tap water	1 litre

Slice the potato and simmer in the water for one hour. Strain and make up the volume to one litre with more tap water. Adjust pH to 6.5. Add the glucose and agar. Heat and mix to dissolve. Sterilize by autoclaving at 121°C for 15 minutes.

6.2.14 Schroth's medium (Schroth *et al.*, 1965)

This medium is a useful selective medium for *Agrobacterium* biovar I.

Base	
mannitol	10.0 g
$NaNO_3$	4.0 g
$MgCl_2.6H_2O$	2.0 g
calcium propionate	1.2 g
$MgHPO_4.3H_2O$	0.2 g
$MgSO_4.7H_2O$	0.1 g
$NaHCO_3$	0.075 g
$MgCO_3$	0.075 g
Oxoid agar No. 3	15.0 g
distilled water	1 litre

Adjust to pH 7.1. Dispense in 95 ml volumes in to bottles and sterilize by autoclaving at 121°C for 15 minutes. When plates are needed, melt the base, cool to 50°C, add the following stock solutions in the amounts stated. Mix and pour plates.

berberine* (0.275 g/10 ml) suspended in sterile distilled water	1.0 ml
sodium selenite (0.1 g/10 ml)	1.0 ml
penicillin G (950 000 units/10 ml)	0.1 ml
streptomycin sulphate or nitrate (0.234 g of base/10 ml)	0.1 ml
actidione (0.224 g active	

*Berberine is insoluble. Do not attempt to sterilize by filtration.

ingredient/10 ml)	1.0 ml
tyrothricin (0.01 g/10 ml)	0.1 ml
bacitracin (6500 units/10 ml)	1.0 ml

Agrobacterium biotype I produces domed, smooth, entire, yellowish- or tan-coloured colonies.

6.2.15 Soil extract agar (see Gordon & Rynearson, 1963) for enhancement of spore production by *Bacillus* spp.

peptone	5.0 g
beef extract	3.0 g
agar (Difco)	15.0 g
tap water } soil extract }	1 litre
pH 7.0	

To prepare soil extract, add 400 g air-dried, sieved garden soil rich in organic matter to 960 ml tap water in a 2 litre conical flask. Sterilize by autoclaving the flask at 121°C for 60 minutes and allow to cool in the unopened autoclave overnight. Filter the extract through paper and re-autoclave the filtrate at 121°C for 20 minutes and allow to stand at room temperature for at least 2 weeks. Use the clear supernatent.

With new batches of soil, make up small amounts of media with varying amounts of soil extract by 10% steps between 10 and 100% vol/vol in the tap water. Use the concentration that gives best spore production by *Bacillus polymyxa* after 3 days.

6.2.16 Starch agar

Oxoid nutrient agar (CM3)	28.0 g
distilled water	1 litre
soluble starch	2.0 g

Dissolve the nutrient agar in the water. Blend the starch in 10 ml water and add to the molten agar. Mix well, dispense into 100 ml volumes and sterilize by autoclaving at 115°C for 10 minutes. Pour plates.

Inoculate and incubate for at least 4 days. Scrape away part of the growth and flood with Lugol's iodine (6.1.3).

The medium turns blue when starch has not been hydrolyzed. A positive reaction is thus given by the appearance of yellowish, clear zones around or under the bacterial growth. Reddish-coloured zones indicate that the starch has been partially hydrolyzed to dextrins.

6.2.17 Sucrose Nutrient Agar

Several green-fluorescent pseudomonads and *Erwinia amylovora* produce large, domed, white-cream, smooth, shining colonies on this medium, probably due to their production of levan, a furfuranose-based polymer. The medium is therefore used for isolation of green-fluorescent pseudomonads and *Erwinia amylovora* and also in the LOPAT scheme for levan production. See also 6.3.9.

Add 5% w/v sucrose to Oxoid nutrient agar CM3. Sterilize by autoclaving at 121°C for 15 minutes.

6.2.18 Sucrose peptone agar (SPA)

Two formulations are given. The first is a general purpose medium which is suitable for growth of *Pseudomonas solanacearum*; the second is a good general medium for fastidious organisms.

6.2.18.1 SPA for *P. solanacearum*

sucrose	20.0 g
peptone	5.0 g
$K_2H\ PO_4$	0.5 g
$MgSO_4.7H_2O$	0.25 g
Oxoid agar No. 3	12.0 g
distilled water	1 litre

Adjust to pH 7.2–7.4. Sterilize by autoclaving at 121°C for 15 minutes.

Freshly-isolated cultures of *Pseudomonas solanacearum* produce flat, fluidal, entire, smooth, white colonies which show internal whorling and streaking.

6.2.18.2 SPA for fastidious bacteria

sucrose	10.0 g
peptone	10.0 g
Oxoid agar No. 3	12.0 g
distilled water	1 litre

Adjust to pH 7.4. Sterilize by autoclaving at 121°C for 15 minutes.

6.2.19 Triphenyl tetrazolium salts (TTC) medium (Kelman 1954)

Virulent colonies of *Pseudomonas solanacearum* and *P. syringae* pv.

phaseolicola can be distinguished from avirulent colonies on this medium.

peptone	10.0 g
casein hydrolysate	1.0 g
glucose	5.0 g
Oxoid agar No. 3	12.0 g
distilled water	1 litre

Adjust pH to 7 approximately. Dispense into 100 ml volumes and sterilize by autoclaving at 121°C for 15 minutes. To the molten agar (55°C) add filter-sterilized 1% aqueous 2:3:5 triphenyl tetrazolium chloride (sodium salt), to give a final concentration of 0.005%

After 2 days virulent wild type colonies of *Pseudomonas solanacearum* are fluidal and either entirely white or with a pale, red centre. Avirulent mutant colonies are butyrous, deep-red often with a bluish border. Virulent *P. s.* pv. *phaseolicola* are red and avirulent strains white.

6.2.20 Yeast dextrose chalk agar

This medium is suitable for isolaton and maintenance of xanthomonads. The chalk buffers the medium. Xanthomonads are not tolerant of acid.

yeast extract or yeastrel	10.0 g
D-glucose	20.0 g
precipated chalk ($CaCO_3$)	20.0 g
Oxoid agar No. 3	12.0 g
distilled water	1 litre

Mix all the ingredients except the chalk and heat to dissolve. Add the chalk. Adjust to pH 7.2 if necessary. Sterilize by autoclaving at 121°C for 15 minutes. Mix well before pouring plates.

Xanthomonads produce large, smooth, domed, mucoid, yellow colonies on this medium.

6.2.21 Yeast glucose mineral agar

A useful medium for fastidious organisms such as *Clavibacter michiganense* subsp. *sepedonicum*.

yeast extract	2.0 g
glucose	2.5 g
K_2HPO_4	0.25 g
KH_2PO_4	0.25 g
$MgSO_4.7H_2O$	0.1 g
$MnSO_4$	0.015 g

NaCl	0.05 g
$FeSO_4.7H_2O$	0.005 g
Oxoid agar No. 3	15.0 g
distilled water	1 litre

Sterilize by autoclaving at 121°C for 15 minutes.

6.2.22 Yeast peptone glucose agar

A medium suitable for isolation and growth of fastidious bacteria such as *Xanthomonas populi* and *Clavibacter michiganense* subsp. *sepedonicum.*

Yeast extract	5.0 g
Oxoid proteose peptone	5.0 g
glucose	10.0 g
Oxoid agar No. 3	12.0 g
distilled water	1 litre

Adjust to pH 6.5–7.0. Autoclave at 115°C for 20 minutes.

6.2.23 Roy and Sasser's medium

A good selective medium for *Agrobacterium tumefaciens* biovar III.

adonitol	4.0 g
boric acid	1.0 g
yeast extract	0.14 g
$MgSO_4$	0.2 g
K_2HPO_4	0.9 g
KH_2PO_4	0.7 g
NaCl	0.2 g
agar	15.0 g
4% aqueous Bravo 500* (fungicide)	0.5 ml
distilled water	1 litre

Adjust pH to 7.2, sterilize by autoclaving at 121°C for 15 minutes, cool to 50°C and add the following after dissolving separately in a few ml of distilled water:

triphenyltetrazolium chloride	80.0 mg
D-cycloserine	20.0 mg

*Available from SDS Biotech, Bayheath House, 4 Fairway, Petswood, Orpington, Kent, BR5 IEG, UK, or Fermenta Plant Protection Co., PO Box 348, 7528 Auburn Rd, Painsville, Ohio 44077, USA.

trimethroprim (with 1 drop IN HCl)	20.0 mg

Colonies of *A. tumefaciens* biovar III appear after 3–4 days at 28°C. They are pearly-white or pinkish with red centres.

6.3 Utilization of carbon compounds

This is a complex topic. Terminology is often confusing with respect to the ability of a bacterium to utilize a particular compound. Assimilation and decomposition effectively imply utilization, but utilization often implies the ability to utilize a compound as the sole source of carbon and of chemical energy.

Utilization may be demonstrated either by the presence of growth or by a biological change brought about in the medium, e.g. acid or alkali shift demonstrated by the use of a suitable indicator. Such tests can also be used to determine whether a bacterium has a respiratory (oxidative) or fermentative metabolism. All known plant pathogenic bacteria except *Clostridium* spp. are aerobic although some, e.g. *Erwinia* spp. are facultative anaerobes and have a fermentative metabolism.

The media used in utilization tests comprise the inorganic minerals essential for growth and where necessary a source of growth factors. The wide variety of such media used by bacteriologists has resulted in divergence of opinion as to the abilities of certain bacteria to utilize compounds as sole carbon and energy sources, although this is not the only reason why such tests have a limited value in identification (1.7). Emphasis on their use to distinguish between related species, subspecies and pathovars has been limited throughout this book to avoid this confusion; more reliance has been placed on other tests which are less likely to yield conflicting results. Nevertheless, utilization tests can be valuable identification tools in particular cases.

6.3.1 Oxidation/Fermentation (O/F) Test (Hugh & Leifson, 1953)

This test is used to indicate whether a particular carbohydrate (usually glucose) is utilized oxidatively or fermentatively by the test bacterium. Carbohydrate utilization resulting in acid formation is determined under aerobic and anaerobic conditions.

Most bacterial plant pathogens are oxidative (respiratory). The erwinias, however, are fermentative.

Base

peptone	2.0 g

NaCl	5.0 g
K_2HPO_4	0.3 g
Oxoid agar No. 3	3.0 g
bromothymol blue	0.03 g (15 ml of 0.2% aqueous solution)
distilled water	1 litre

Adjust to pH 7.1. Distribute in convenient volumes (e.g. 90 ml) and sterilize by autoclaving at 121°C for 15 minutes.

To the molten base (50°C) add, for example, 10 mls of 10% aqueous D-glucose solution (filter-sterilized) to obtain a final concentration of 1%. Dispense in 6 ml volumes in sterile, narrow tubes.

Stab-inoculate two tubes of the medium and seal one tube with 1–2 cm depth of sterile vaspar (6.8) or liquid paraffin. Incubate for 3 days (coryneforms may require longer incubation).

The production of a yellow colour (indicating acid production) in the unsealed tube but not the sealed tube indicates oxidative metabolism of glucose. Acid in both tubes indicates fermentative metabolism. Gas production, if it occurs, will be seen in the sealed tube. Absence of acid in both tubes indicates that glucose is not utilized under the conditions of the test or that alkaline breakdown products of peptone degradation buffered weak acid production from the carbohydrate. In such cases, which are rare, repeat the test in a medium using the inorganic salts of Ayers, Rupp and Johnson base (C2) plus 0.05% yeast extract instead of the peptone, NaCl and K_2HPO_4 of this medium. Very rarely, bacteria are inhibited by bromothymol blue, and if this is suspected the indicator can be added after incubation and growth.

6.3.2 Ayers, Rupp and Johnson base (carbohydrate and organic acid utilization base) (Ayers, Rupp & Johnson, 1919)

This is a general purpose synthetic base containing no organic nutrients and may be used for carbohydrate and organic acid utilization tests for many plant pathogenic bacteria, particularly the non-fluorescent and fluorescent pseudomonads. More appropriate bases are listed for other genera.

Base

$NH_4H_2PO_4$	1.0 g
KCl	0.2 g

$MgSO_4.7H_2O$	0.2 g
Oxoid Agar No. 3 (optional)	12.0 g
distilled water	1 litre

Adjust to pH 7.0–7.2 if necessary. Distribute base into known convenient amounts and sterilize by autoclaving at 121°C for 15 minutes. For demonstration of acid production incorporate 0.0016% (w/v) bromocresol purple.

Carbohydrate solutions

Prepare filter-sterilized solutions of test carbohydrates (neutralize if necessary).

Add to the sterile base (molten, if necessary, at 45°C) to obtain the following concentrations:

1 Non-fluorescent pseudomonads — liquid base; sugars at 0.2%, other carbohydrates at 0.1%.

2 Fluorescent pseudomonads — agar base; sugars at 0.5%, other carbohydrates at 0.1%.

Record presence or absence of growth after appropriate length of incubation.

6.3.3 Acid from sucrose test

This is used as a secondary test in the LOPAT grouping for plant-pathogenic green-fluorescent pseudomonads.

Base

Prepare Ayers, Rupp and Johnson agar base (6.3.2), containing 0.0016% bromocresol purple. Distribute in known convenient volumes and sterilize by autoclaving at 121°C for 15 minutes.

Sucrose solution

Prepare 10% (w/v) aqueous sucrose solution and filter-sterilize. Add to the molten base at 45°C to a final concentration of 1%. Allow to set as agar slants.

Record presence or absence of a yellow colour, indicating acid production, within 3 days.

6.3.4 Purple peptone base (Dye, 1968) (acid production)

This medium is useful for determining acid production from carbohydrate breakdown by all *Erwinia* spp.

Base

peptone	10.0 g
1.5% alcoholic bromocresol purple	0.7 ml
distilled water	1 litre

Adjust to pH 6.8. Dispense base into known, convenient volumes and sterilize by autoclaving at 121°C.

Carbohydrate solutions

Prepare 10% aqueous solutions of the test solutions and filter-sterilize. Add test sugars to the base to a final concentration of 1%.

Record presence or absence of a yellow colour, indicating acid production, within 14 days.

6.3.5 Biovar Medium for Pseudomonas solanacearum (Hayward, 1964)

This medium is used for distinguishing the four biovars of *Pseudomonas solanacearum* and is based upon utilization of various carbohydrates.

Base

$NH_4H_2PO_4$	1.0 g
KCl	0.2 g
$MgSO_4.7H_2O$	0.2 g
peptone	1.0 g
1.% (w/v) aqueous bromothymol blue	0.3 ml
Oxoid Agar No. 3	1.5 g
distilled water	1 litre

Adjust pH to 7.1. Dispense in 4.5 ml amounts in test tubes and sterilize by autoclaving at 121°C for 15 minutes.

Carbohydrate solutions

Prepare 10% aqueous solutions of maltose, lactose, cellobiose, mannitol, sorbitol and dulcitol. Sterilize dulcitol by autoclaving at 115°C for 15 minutes. Sterilize the other carbohydrates by filtration.

To the molten base (45°C), add 0.5 ml carbohydrate solution to each tube. Allow to cool.

Inoculate one tube of each medium plus one tube of base. Record presence or absence of a yellow coloration, indicating acid production, within 14 days. Read against a white background in good light since the colour change from green to yellow may be difficult to detect. Reactions of the four biovars are:

	Biovar			
Carbohydrate source	I	II	III	IV
Maltose	−	+	+	−
Lactose	−	+	+	−
Cellobiose	−	+	+	−
Mannitol	−	−	+	+
Sorbitol	−	−	+	+
Dulcitol	−	−	+	+

6.3.6 Gluconate medium for 2-ketogluconate production

Medium for fluorescent pseudomonads

tryptone	1.5 g
yeast extract	1.0 g
K_2HPO_4	1.0 g
sodium gluconate	40.0 g
distilled water	1 litre

Adjust to pH 7.0 if necessary. Dispense in 5 ml volumes into capped tubes. Sterilize by autoclaving at 121°C for 15 minutes.

Test method. Inoculate and grow in shake culture for 3 days.

Add an equal volume of Benedict's quantitative reagent to each tube, mix and heat in a boiling water bath for 10 minutes. Use an uninoculated tube as control. A positive is given by the formation of a brown or orange precipitate.

6.3.7 3-ketolactose test

This test usefully differentiates the common biovar I of the genus *Agrobacterium* from the other biovars.

Medium

lactose	10.0 g
yeast extract	1.0 g

Oxoid agar No. 3	12.0 g
distilled water	1 litre

Sterilize by autoclaving at 121°C for 15 minutes. Pour plates.

Test method. Inoculate a spot heavily using a log phase nutrient dextrose agar culture. Incubate for 2 days and flood with Benedict's qualitative reagent.

Record presence or absence of a yellow zone of Cu_2O, indicating 3-ketolactose production, in the agar around the colony within 1 hour. Biovar I produces this compound.

6.3.8 Reducing compounds from sucrose

This test is used to distinguish between *Erwinia carotovora* subspp. *carotovora* and *atroseptica*.

Medium

sucrose	40.0 g
peptone	10.0 g
Lab-lemco beef extract	5.0 g
distilled water	1 litre

Adjust to pH 7 approximately. Dispense in 5 ml volumes into tubes and sterilize by autoclaving at 121°C for 15 minutes.

Inoculate and incubate on a shaker for 2 days. Add an equal volume of Benedict's quantitative reagent to the test tube and to an uninoculated control tube, shake and place in boiling water for 10 minutes. The formation of an orange or brown precipitate in the inoculated tube indicates the presence of reducing compounds.

6.3.9 Levan production from sucrose

Levan is a polymer synthesized from the fructose moiety of the sucrose molecule, and its formation is usually responsible for the production of mucoid colonies by some *Pseudomonas* and *Erwinia* spp. on media high in sucrose (6.2.17). It should be noted that in this text 'levan-type colonies' refers only to this type of characteristic colony and does not imply that all the bacteria involved have been shown to produce levan, although it is highly likely that they do.

Levan-type colonies are pearly-white, shining, high-domed and mucoid. They are semi-transparent to tranlucent initially, but after 2–3 days become translucent to semi-opaque. Old colonies may collapse, probably through the production by the bacteria of levanase.

This colony form is a very useful diagnostic character, but until experienced, it is advisable to run parallel plates of pure cultures of known levan producers such as *Erwinia amylovora*, *Pseudomonas syringae* pv. *syringae* and *P. marginalis*. All three have characteristic but different levan-type colonies that, with practice, can be identified with reasonable certainty on isolation plates.

6.4 Utilization of nitrogen compounds

6.4.1 Nitrate reduction test

Some oxidative bacteria, including many pseudomonads, have the ability to grow in anaerobic conditions rich in nitrate ions. Energy is derived by reducing nitrate to nitrite by means of nitrate reductase coupled to an electron transport chain.

Since nitrite is toxic some bacteria further reduce nitrite to nitrous oxide and/or gaseous nitrogen by the nitrite reductase system. This denitrification process is often referred to as nitrate respiration. Some bacteria may reduce nitrate to nitrite under aerobic conditions.

Tests for nitrate reduction involve growing the test bacteria in a nitrate medium and then testing for the presence of nitrite using Follet and Ratcliff's reagent (6.4.1.2)*. If no nitrite can be detected, either nitrate has not been reduced or the nitrate has been reduced to nitrite which has been further reduced. If zinc or cadmium dust is now added to the mixture of culture and Follet and Ratcliff's reagent, any nitrate present will be reduced by the metal to nitrite and react with the reagent. If there is then no reaction, this indicates that both nitrate and nitrite are absent from the medium and that denitrification has occurred.

6.4.1.1 **Nitrate medium** (Stanier *et al.*, 1966)

KNO_3	1.0 g
peptone	5.0 g
yeast extract	3.0 g
Oxoid agar No. 3	3.0 g
distilled water	1 litre

Adjust to pH 7–7.2 if necessary. Dispense into 8 ml volumes in test tubes. Sterilize by autoclaving at 121°C for 15 minutes.

*An old method of nitrite detection with α-napthylamine is no longer recommended because of its potential health hazard.

6.4.1.2 Follet and Ratcliff's reagent

Glacial acetic acid	50.0 ml
distilled water	360.0 ml
sulphanilic acid	0.25 g
α-naphthol	0.2 g
10% (v/v) aqueous 0.88 ammonia solution	90.0 ml

Add the acetic acid to the water, mix, warm to 50°C and pour into a dark, glass bottle containing the sulphanilic acid. When this has dissolved add the α-naphthol and allow this to dissolve. Cool the mixture to room temperature and then add the ammonia solution. Store in the dark at 4°C.

6.4.1.3 Test method

Stab inoculate the selected medium in duplicate and incubate for 3–5 days, until good growth has been obtained. Add an approximately equal volume of Follet and Ratcliff's reagent to the contents of one tube and an uninoculated control at 3 and 5 days. Shake well.

The development of an orange to orange-brown colour is a positive reaction and indicates the presence of nitrite. If the colour does not develop within 10 minutes, add a knife point of cadmium powder and record the reduction of nitrate beyond nitrite if no orange to orange-brown colour develops in 10 minutes — if this colour does develop record absence of nitrate reduction. An uninoculated control should develop the colour in the presence of reagent and the metal.

6.4.2 Urease activity

6.4.2.1 Urease medium (Christensen's)

Base

neutralized peptone (Oxoid L34)	1.0 g
D-glucose	1.0 g
NaCl	5.0 g
KH_2PO_4	2.0 g
1.2% aqueous phenol red solution	1.0 ml
Oxoid agar No. 3	12.0 g
distilled water	1 litre

Adjust to pH 6.8 if necessary. Dispense in 90 ml volumes and sterilize by autoclaving at 121°C for 15 minutes. Filter-sterilize a 20% aqueous urea solution. To the molten base (55°C) add the sterile urea solution to a final concentration of 2%. Mix well and dispense in 6 ml volumes into sterile test tubes and allow to set as slopes.

6.4.2.2 Test method

Inoculate and incubate for up to 7 days and observe for urease activity, indicated by the development of a pink-red coloration, as soon as good growth is obtained.

6.4.3 Arginine dihydrolase test (Thornley, 1960)

Medium

neutralized peptone (Oxoid L34)	1.0 g
NaCl	5.0 g
K_2HPO_4	0.3 g
L (+) arginine HCl	10.0 g
phenol red	0.01 g
agar	3.0 g
distilled water	1 litre

Mix, dissolve by heating, sterilize by autoclaving at 121°C for 15 minutes. Adjust pH to 7.2. Dispense 3 ml volumes in bijou bottles.

Test method. Stab inoculate and seal by adding approximately 2 ml Vaspar (6.8). Incubate for 3 days and record a positive reaction if there is a strong alkaline shift, indicated by a colour change from an orange tint to a definite cherry red. An alkaline shift may sometimes occur during autoclaving if pH is adjusted prior to autoclaving, hence the need to adjust afterwards.

6.4.4 Indole test

Medium

Dispense a 1% aqueous solution of tryptone plus 0.01% tryptophane into tubes in 5 ml volumes. Sterilize by autoclaving at 121°C for 15 minutes.

Kovacs' indole reagent

p-dimethylaminobenzaldehyde	5.0 g
amyl alcohol	75.0 ml
conc. HCl	25.0 ml

Dissolve the aldehyde in the alcohol by heating gently in a water bath at 50°C. Cool and add the acid to give a straw-brown coloured liquid. Store in a dark glass bottle at 4°C.

Test method. Inoculate two tubes of the appropriate medium and incubate on a shaker. Test for indole after 2 and 5 days by adding 0.5 ml of reagent to a tube. The appearance of a red colour in the reagent layer within one minute indicates the presence of α-methyl indole.

6.5 Degradation of macromolecules

6.5.1 Egg yolk medium for lecithinase production

Medium

Add 10 ml of a sterile aqueous egg yolk emulsion (Oxoid SR37) to 100 ml of Oxoid nutrient agar (CM3) at 45°C. Pour plates.

Test method. Spot cultures on to the medium and incubate for 3 days.

Positive reaction: lecithinase production is shown by the production of opalescence around the colonies or the formation of a pearly layer over and around the colonies. Alternatively, flood the plate with saturated copper sulphate solution and look for an insoluble, bright greenish-blue precipitate.

6.5.2 Gelatin hydrolysis

Medium

yeast extract	3.0 g
peptone	5.0 g
gelatin	120.0 g
distilled water	1 litre

Adjust pH to approximately 7 if necessary, after allowing the solids to stand in the water for 15 minutes followed by heating to dissolve.

Dispense into tubes to a depth of approximately 5 cm. Sterilize by autoclaving at 121°C for 15 minutes.

Test method. Stab inoculate; also use an uninoculated control. Incubate and record liquefaction of the medium. It may be necessary to cool the tubes together with an uninoculated control at 5°C for 30 minutes before recording the results.

6.5.3 Starch hydrolysis

Medium

nutrient agar	28.0 g
distilled water	1 litre
soluble starch	2.0 g

Dissolve the nutrient agar powder in the water by heating. Dissolve the starch in 10 ml distilled water and add to the molten agar. Mix well, dispense in 100 ml volumes and sterilize by autoclaving at 115°C for 10 minutes. Pour plates.

Test method. Inoculate and incubate for at least 7 days unless stated otherwise. Scrape away some of the growth and flood the plate with Lugol's iodine (6.1.3) and record presence (or absence) of starch hydrolysis, indicated by the presence of clear zones in the black-stained medium around or under the colonies. Reddish zones indicate that starch has been partially hydrolyzed to dextrins, and is usually considered a negative reaction.

6.5.4 Aesculin medium

peptone (Oxoid L37)	10.0 g
aesculin	1.0 g
ferric citrate scales	0.5 g
agar	12 g
distilled water	1 litre

Mix, heat to dissolve, dispense in 5 ml volumes into tubes and sterilize by autoclaving at 115°C for 15 minutes. Allow to set as slopes. For some tests arbutin is substituted for aesculin.

Test method. Streak culture on to slope. Aesculin utilization is indicated by the development within 3–4 days of a progressive blackening of the medium or disappearance of fluorescence when observed under ultra-violet light.

6.5.5 Tween 80 agar

Base

peptone	10.0 g

NaCl	5.0 g
Ca $Cl_2.2H_2O$	0.1 g
agar	15.0 g
water	1 litre

Adjust to pH 7.2–7.4. Sterilize the base and Tween 80 separately by autoclaving at 121°C for 15 minutes. Add Tween 80 to the molten base to give a final concentration of 1%, mix well and pour plates. Incubate for up to 7 days. Development of a milky-white precipitate around the colonies demonstrates the ability to hydrolyze the lipid Tween 80.

6.6 Miscellaneous tests

6.6.1 Catalase test

Add a few drops of 3% hydrogen peroxide to growth on any non-blood-containing medium. Look for gas bubbles immediately and if absent look again after 5 minutes. Record as catalase-positive if bubbles of gas form.

6.6.2 Oxidase test (Kovacs, 1956)

This test is used to detect the presence of cytochrome C oxidase (oxidase positive). It is essential to use a platinum loop rather than a nichrome loop.

Reagent

Weigh 0.01 g samples of tetramethylphenylene diamine dihydrochloride into clean bijou bottles, cap tightly and seal with Parafilm. Store at room temperature.

Test method. Add 1 ml of distilled water to give a 1% solution (always use freshly-prepared solutions). Pipette a few drops of the reagent on to filter paper. Using a platinum loop, rub a heavy loopful of culture from a 1–2 day nutrient agar across the impregnated filter paper and record the development of a purple colour within 10 seconds as a positive reaction.

Reactions occurring between 10 and 30 seconds may be regarded as weekly positive. Avoid testing growth from media rich in sugars if possible since this may give false positive results. If the test bacterium will not grow without added carbohydrate use the minimum concentration that will give growth.

6.6.3 Phosphatase test

Medium

Prepare Oxoid Nutrient Agar (CM3), distribute in 100 ml volumes and sterilize by autoclaving at 121°C for 15 minutes. To 100 ml molten medium at 50°C add 1 ml filter-sterilized 5% aqueous disodium phenolphthalein diphosphate solution to give a final concentration of 0.05%. Mix well and pour plates. Streak or spot medium to inoculate and incubate for 2 days.

Test method. Add a few drops of 0.88 ammonia solution to the lid of the Petri dish and invert the medium over it; record development of a pink colour in the bacterial growth as a positive result.

6.6.4 Potato plugs

Scrub large potato tubers thoroughly with soap and water. Rinse and take cores approximately 5 cm long with a 1 cm cork borer. Place immediately into distilled water. Remove the peel, cut one end diagonally to make a slope at approximately 45°C. Place plug in a tube, diagonal end uppermost and add sufficient distilled water to cover half the plug. For *Erwinia salicis*, 2% aqueous glycerol to enhance pigment production may be used instead. Cap tubes and sterilize by autoclaving at 115°C for 10 minutes.

Test method. Inoculate the slope and incubate for approximately 3 days.

Xanthomonas campestris pathovars produce a dense, spreading, yellow honey-coloured growth. The potato tissue is softened and the edges become rounded. *Erwinia salicis* produces a yellow pigment, though older cultures may lose this character.

6.6.5 Tetrazolium salt (TTC) tolerance tests

Xanthomonads are not readily distinguished from other yellow, oxidative Gram-negative rods by nutritional or physiological tests. They are, however, unusually sensitive to tetrazolium salts.

Medium

Prepare nutrient agar (Oxoid CM3) in 100 ml volumes and sterilize by autoclaving at 121°C for 15 minutes. To the molten agar at 55°C add filter-sterilized 1% aqueous 2:3:5 triphenyl tetrazolium chloride (TTC) to give two final TTC concentrations: 0.02% and 0.1%. Pour plates.

Test method. Inoculate by spotting plates of each TTC concentration together with a nutrient agar control. Record presence or absence of growth. Four cultures can be tested on each plate.

Most yellow-pigmented xanthomonads are inhibited by 0.02% TTC. All are inhibited by 0.1% TTC.

6.6.6 Salt tolerance test

Prepare nutrient broth and add sodium chloride to the required concentration (w/v), e.g. 2% for distinguishing between *Agrobacterium* biovars and 7% for distinguishing between the *Curtobacterium flaccumfaciens* and *Clavibacter michiganense* groups. Sterilize by autoclaving at 121°C for 15 minutes.

Inoculate and incubate on a shaker for up to 5 days or as recommended. Bacterial growth seen as turbidity signifies NaCl tolerance. Always use a control with no added sodium chloride. Some commercial brands of nutrient broth already contain some sodium chloride. The final concentration should take this into account.

6.6.7 Pasteurization of endospores

Bacillus spp. often isolated from soft-rotted tissues do not always produce large numbers of endospores and a spore stain may be of little value. Pasteurization aims to kill all the vegetative cells allowing only endospores (however few) to survive and germinate on further incubation. Heating to sub-lethal doses tends to age spores and enhance germination.

Heat turbid bacterial suspensions contained in thin-walled sterile tubes in a water bath at a temperature of 80°C for 30 minutes taking usual sterile precautions and immersing tubes deeply. Pipette about 1 ml of heated suspension into molten (*c.* 50°C) nutrient agar and pour into a Petri dish, or mix enough heated suspension with nutrient broth to give a faintly turbid suspension. Incubate and observe for growth which signifies thermoduric qualities usually indicative of endospore formation.

6.7 **Buffers**

All buffers can be autoclaved. pH should be checked after preparation.

6.7.1 Phosphate (0.05 mol/l pH 7.02)

Na_2HPO_4	4.26 g
KH_2PO_4	2.72 g
distilled water	1 litre

6.7.2 Phosphate-buffered saline (0.01 mol/l pH 7.2)

NaCl	8.0 g
$Na_2HPO_4.12H_2O$	2.7 g
$NaH_2PO_4.2H_2O$	0.4 g
distilled water	1 litre

6.7.3 Phosphate-buffered saline (0.01 mol/l pH 7.6)

NaCl	8.0 g
$Na_2HPO_4.12H_2O$	3.227 g
$NaH_2PO_4.2H_2O$	0.156 g
distilled water	1 litre

6.7.4 Phosphate–glycerine (0.1 mol/l pH 7.6)

This buffer is used as the mountant for fluorescent antibody tests.

$Na_2HPO_4.12H_2O$	3.2 g
NaH_2PO_4	0.15 g
glycerol	50.0 ml
distilled water	100.0 ml

6.7.5 TRIS buffer (0.1 mol/l pH 7.4)

0.2 mol/l TRIS (hydroxymethyl)-methylamine	50.0 ml
0.2 mol/l HCl	41.4 ml
Mix and make up to 100 ml	

6.8 **Vaspar**

1 part white petroleum jelly (e.g. Vaseline)
3 parts liquid paraffin

Melt the Vaseline and add the liquid paraffin. Mix well, dispense into small (20 ml) bottles and sterilize by autoclaving at 121°C for 15 minutes.

References

Anderson, A.R. & Moore, L.W. (1979) Host specificity in the genus Agrobacterium. *Phytopathology* **69**, 320–3.

Ark, P.A. & Thompson, J.P. (1961) Detection of hairy root pathogen, *Agrobacterium rhizogenes* by the use of fleshy roots. *Phytopathology* **51**, 69–71.

Ayers, S.H., Rupp, P. & Johnson, W.T. (1919) A study of the alkali-forming bacteria in milk. *United States Department of Agriculture Bulletin* **782**.

Barzic, M.R. & Trigalet, A. (1982) Détection de Pseudomonas phaseolicola (Burkh.) Dowson par la technique ELISA. *Agronomie* **2**, 389–98.

Bazzi, C. (1979) Identification of *Pseudomonas cepacia* on onion bulbs in Italy. *Phytopathologische Zeitschrift* **95**, 254–8.

Beijer, J.J. (1972) Course of yellow disease caused by *Xanthomonas hyacinthi* (Wakk.) Dowson in the leaf and bulb of hyacinth. *Mededeelingen van de Landbouwhoogeschool, Wageningen* **72** (**30**), 1–52.

Billing, E. (1970) *Pseudomonas viridiflava* (Burkholder, 1930) Clara, 1934, *Journal of Applied Bacteriology* **33**, 492–500.

Bird, A.F. (1981) The *Anguina–Corynebacterium* association. In *Plant Parasitic Nematodes*, Vol. III, pp. 303–23. (Ed B.M. Zuckerman & R.A. Rohde). Academic Press, New York.

Bird, A.F. & Stynes, B.A. (1977) The morphology of a *Corynebacterium* sp. parasitic on annual rye grass. *Phytopathology* **67**, 828–30.

Bird, L.S., Brinkerhoff, L.A. & Davis, R.G. (1981) Bacterial blight. In *Compendium of Cotton Diseases*, pp. 25–8. (Ed G.M. Watkins). American Phytopathological Society, St Paul, Minnesota.

Boa, E.A. & Preece, T.F. (1981) *Pseudomonas savastanoi* var. *fraxini*: symptoms of ash canker disease in the field in Scotland and England. *Proceedings of the 4th International Conference on Plant Pathological Bacteria (Angers)* **2**, 917–28.

Bradbury, J.F. (1967a) *Erwinia stewartii. C.M.I. Descriptions of Pathogenic Fungi and Bacteria* **123**. Commonwealth Mycological Institute, Kew.

Bradbury, J.F. (1967b) *Pseudomonas lachrymans. C.M.I. Descriptions of Pathogenic Fungi and Bacteria* **124**. Commonwealth Mycological Institute, Kew.

Bradbury, J.F. (1967c) *Pseudomonas tabaci. C.M.I. Descriptions of Pathogenic Fungi and Bacteria* **129**. Commonwealth Mycological Institute, Kew.

Bradbury, J.F. (1967d) *Xanthomonas juglandis. C.M.I. Descriptions of Pathogenic Fungi and Bacteria* **130**. Commonwealth Mycological Institute, Kew.

Bradbury, J.F. (1970a) *Aplanobacter populi. C.M.I. Descriptions of Pathogenic Fungi and Bacteria* **231**. Commonwealth Mycological Institute, Kew.

Bradbury, J.F. (1970b) *Erwinia tracheiphila. C.M.I. Descriptions of Pathogenic Fungi and Bacteria* **233**. Commonwealth Mycological Institute, Kew.

Bradbury, J.F. (1970c) *Xanthomonas oryzae. C.M.I. Descriptions of Pathogenic Fungi and Bacteria* **239**. Commonwealth Mycological Institute, Kew.

Bradbury, J.F. (1970d) *Xanthomonas oryzicola. C.M.I. Descriptions of Pathogenic Fungi and Bacteria* **240**. Commonwealth Mycological Institute, Kew.
Bradbury, J.F. (1973a) *Pseudomonas avenae. C.M.I. Descriptions of Pathogenic Fungi and Bacteria* **371**. Commonwealth Mycological Institute, Kew.
Bradbury, J.F. (1973b) *Pseudomonas caryophylli. C.M.I. Descriptions of Pathogenic Fungi and Bacteria* **373**. Commonwealth Mycological Institute, Kew.
Bradbury, J.F. (1973c) *Corynebacterium tritici. C.M.I. Descriptions of Pathogenic Fungi and Bacteria* **377**. Commonwealth Mycological Institute, Kew.
Bradbury, J.F. (1973d) *Xanthomonas ampelina. C.M.I. Descriptions of Pathogenic Fungi and Bacteria* **378**. Commonwealth Mycological Institute, Kew.
Bradbury, J.F. (1973e) *Xanthomonas vasculorum. C.M.I. Descriptions of Pathogenic Fungi and Bacteria* **380**. Commonwealth Mycological Institute, Kew.
Bradbury, J.F. (1977) *Xanthomonas fragariae. C.M.I. Descriptions of Pathogenic Fungi and Bacteria* **558**. Commonwealth Mycological Institute, Kew.
Brisbane, P.G. & Kerr, A. (1983) Selective media for three biovars of *Agrobacterium. Journal of Applied Bacteriology* **54**, 425–31.
Brocklehurst, T.F. & Lund, B.M. (1981) Properties of pseudomonads causing spoilage of vegetables stored at low temperature. *Journal of Applied Bacteriology* **50**, 259–66.
Brown, N.A. (1932) Canker of ash trees, produced by a variety of the olive-tubercle organism, *Bacterium savastanoi. Journal of Agricultural Research* **44**, 701–22.
Bryan, M.K. (1928) Lilac blight in the United States. *Journal of Agricultural Research* **36**, 225.
Buchanan, R.E. & Gibbons, N.E. (1974) *Bergey's Manual of Determinative Bacteriology*, 8th Edn. Williams & Wilkins, Baltimore.
Buddenhagen, I.W. (1961) Bacterial wilt of bananas: history and known distribution. *Tropical Agriculture (Trinidad)* **38**, 107–21.
Buddenhagen, I.W. & Kelman, A. (1964) Biological and physiological aspects of bacterial wilt caused by *Pseudomonas solanacearum. Annual Review of Phytopathology* **2**, 201–30.
Burdekin, D. (1972) Bacterial canker of poplar. *Annals of Applied Biology* **72**, 295–9.
Burkholder, W.H. (1950). Sour skin, a bacterial rot of onion bulbs. *Phytopathology* **40**, 115–17.
Chaldecott, M.A. & Preece, T.F. (1983) The use of a tomato cotyledon test to identify *Corynebacterium michiganense* pv. *michiganense. Plant Pathology* **32**, 441–8.
Chatigny, M.A. (1961) Protection against infection in the microbiology laboratory. Devices and procedures. *Advances in Applied Microbiology* **3**, 131–92.
Christensen, P. (1977) Synonymy of *Flavobacterium pectinovorum* Dorey with *Cytophaga johnsonae* Stanier. *International Journal of Systematic Bacteriology* **27**, 122–32.
Christopher, W.N. & Edgerton, C.W. (1930) Bacterial stripe diseases of sugarcane in Louisiana. *Journal of Agricultural Research* **41**, 259–67.
Cirulli, M. & Ercolani, G.L. (1969) Identificazione di *Pseudomonas pisi* Sackett in Italia. *Phytopathologica Mediterranea* **8**, 234–7.
Clark, A.G. (1969) A selective medium for the isolation of *Agrobacterium* species. *Journal of Applied Bacteriology* **32**, 348–51.
Clark, M.F. (1981) Immunosorbent assays in Plant Pathology. *Annual Review of Phytopathology* **19**, 83–106.
Clark, M.F. & Adams, A.N. (1977) Characteristics of the microplate method of enzyme-linked immunosorbent assay for the detection of plant viruses. *Journal of General Virology* **34**, 475–83.
Close, R. & Mulcock, A.P. (1972) Bacterial wilt, *Corynebacterium insidiosum* (McCulloch, 1925) Jensen 1934 of lucerne in New Zealand. *New Zealand Journal of Agricultural Research* **15**, 141–8.

Collins, C.H., Hartley, E.G. & Pillsworth, R. (1974) *The prevention of laboratory acquired infections. Public Health Laboratory Service Monograph Series* No. 6, Her Majesty's Stationery Office, London.

Crosse, J.E. (1956) An epidemic leaf spot and spur wilt of cherry caused by *Pseudomonas morsprunorum. Report of East Malling Research Station, 1955*, 121–5.

Crosse, J.E. (1957) Bacterial canker of stone-fruits. III. Inoculum concentration and time of inoculation in relation to leaf-scar infection of cherry. *Annals of Applied Biology* **45**, 19–35.

Crosse, J.E. (1966) Epidemiological relations of the pseudomonad pathogens of deciduous fruit trees. *Annual Review of Phytopathology* **4**, 291–310.

Crosse, J.E. & Garrett, C.M.E. (1966) Bacterial canker of stone-fruits. VII. Infection experiments with *Pseudomonas morsprunorum* and *P. syringae*. *Annals of Applied Biology* **58**, 31–41.

Cuppels, D. (1984) Phages useful in detecting and identifying *Pseudomonas syringae* pv. *tomato*. *Proceedings of the 2nd Working Group on Pseudomonas syringae pathovars. Sounion, Greece, April 1984*, pp. 50–2. (Ed C.G. Panagopoulos, P.G. Psallidas & A.S. Alivizatos). Hellenic Phytopathological Society, Athens.

Cuppels, D. & Kelman, A. (1974) Evaluation of selective media for isolation of soft-rot bacteria from soil and plant tissue. *Phytopathology* **64**, 468–75.

Cutri, S.S., Maculey, B.J. & Roberts, W.P. (1984) Characteristics of pathogenic non-fluorescent (smooth) and non-pathogenic fluorescent (rough) forms of *Pseudomonas tolaasii* and *Pseudomonas gingeri*. *Journal of Applied Bacteriology*, **57**, 291-8.

Daniel, J.F., Gardan, L., Luisetti, J. & Prunier, J.P. (1975) Identification de l'agent responsable de la maladie des taches noires de la mangue (*Mangifera indicae*). *Fruits* **30**, 625–9.

Davis, R.G. & Sandridge, T.L. (1977) Epidemiology of bacterial blight of cotton. *Mississippi Agricultural and Forestry Experiment Station Technical Bulletin* **88**.

DeBoer, S.H., Copeman, R.J. & Vruggink, H. (1979) Serogroups of *Erwinia carotovora* potato strains determined by diffusible somatic antigens. *Phytopathology* **69**, 316–19.

De Cleene, M. & De Ley, J. (1976) The host range of crown gall. *Botanical Review* **42**, 389–466.

De Cleene, M. & De Ley, J. (1981) The host range of infectious hairy root. *Botanical Review* **47**, 147–94.

De Cleene, M., Leyns, F., van den Mooter, M., Swings, J. & De Ley, J. (1981) Reaction of grass varieties grown in Belgium to *Xanthomonas campestris* pv. *graminis*. *Parasitica* **37**, 29–34.

De Kam, M. (1980) Three xanthomonads on poplar and willow in the Netherlands. *Proceedings of the FAO/IUFRO Symposium on Resistance mechanisms in Poplar*, Poland, pp. 169–71.

Department of Health and Social Security (1972) *Safety in Pathology Laboratories*. Her Majesty's, Stationery Office, London.

DeVay, J.E. & Schnathorst, W.C. (1963) Single-cell isolation and preservation of bacterial cells. *Nature* **199**, 775–7.

Dickey, R.S. (1979) *Erwinia chrysanthemi*: A comparative study of phenotypic properties of strains from several hosts and other *Erwinia* species. *Phytopathology* **69**, 324–9.

Dickey, R.S. (1981) *Erwinia chrysanthemi*: Reaction of eight plant species to strains from several hosts and to strains of other *Erwinia* species. *Phytopathology* **71**, 23–9.

Dimock, A.W. (1950) Major carnation diseases and their spread. *Florists Exchange, New York* **114**, 21 and 72.

Dinesen, I.G. (1980) Bacterial wilt of beans. *Danish Journal of Plant and Soil Science* **84**, 361–70.

Doidge, E.M. (1915) The South African Mulberry Blight, *Bacterium Mori* (Boy. and Lamb.) Smith. *Annals of Applied Biology* **2**, 113.

Dowson, W.J. (1957) *Plant Diseases Due to Bacteria.* Cambridge University Press, Cambridge.

Dye, D.W. (1963) *Xanthomonas begoniae* (Takimoto 1934), Dowson 1939, in New Zealand. *New Zealand Journal of Science* **6**, 313–19.

Dye, D.W. (1968) A taxonomic study of the genus *Erwinia.* 1. The 'amylovora' group. *New Zealand Journal of Science* **11**, 590–607.

Dye, D.W. Bradbury, J.F., Goto, M., Hayward, A.C., Lelliott, R.A. & Schroth, M.N. (1980) International standards for naming pathovars of phytopathogenic bacteria and a list of pathovar names and pathotype strains. *Review of Plant Pathology* **59**, 153–68.

Dye, D.W. & Wilkie, J.P. (1973) Angular leaf spot of strawberry in New Zealand. *New Zealand Journal of Agricultural Research* **16**, 311–14.

Ebbels, D.L. (1976) Diseases of upland cotton in Africa. *Review of Plant Pathology* **55**, 747–63.

Egli, T., Goto. M. & Schmidt, D. (1975) Bacterial wilt, a new forage grass disease. *Phytopathologische Zeitschrift* **82**, 111–21.

Egli, T. & Schmidt, D. (1982) Pathogenic variation among the causal agents of bacterial wilt of forage grasses. *Phytopathologische Zeitschrift* **104**, 138–50.

Elliott, C. (1951) *Manual of Bacterial Plant Pathogens, 2nd Edn.* Chronica Botanica, Waltham, Massachusetts.

Emmett Barkley, W. (1981) Containment and disinfection. In *Manual of Methods for General Bacteriology*, (Ed P. Gerhardt, R.G.E. Murray, R.N. Cotislow, E.W. Nester, W.A. Wood, N.R. Krieg & G. Briggs Phillips). American Society for Microbiology, Washington D.C.

EPPO (European and Mediterranean Plant Protection Organization) (1961) *Report of International Conference on Corynebacterium sepedonicum and Pseudomonas solanacearum, Paris, 1960.* EPPO Publications, Series A, No. 31.

Ercolani, G.L. (1970) Individuazione di *Xanthomonas pruni* (E.F. Smith) Dowson quale agente della maculatura batterica del susino Emilia. *Phytopathologica Mediterranea* **9**, 62-4.

Ercolani, G.L. (1983) Variability among isolates of *Pseudomonas syringae* pv. *savastanoi* from the phylloplane of the olive. *Journal of General Microbiology* **129**, 901–16.

Fahy, P.C., Nair, N.G. & Bradley, J.K. (1981) Epidemiology and biological control of bacterial blotch caused by *Pseudomonas tolaasii. Mushroom Science* **11**, 343–52.

Fahy, P.C. & Persley, G.J. (1983) *Plant Bacterial Diseases: A Diagnostic Guide.* Academic Press, London.

Fang, C.T., Ren, H.C., Chen, T.Y., Chu, Y.K., Faan, H.C. & Wu, S.C. (1957) A comparison of the rice bacterial leaf blight organism with the bacterial leaf streak organism of rice and *Leersia hexandra* Swartz. *Acta Phytopathologica, Sinica* **3**, 99–124.

Feltham, R.K.A., Power, A.K., Pell, P.A. & Sneath, P.H.A. (1978) A simple method for storage of bacteria at −76°C. *Journal of Applied Bacteriology* **44**, 313–16.

Fuerst, J.A. & Hayward, A.C. (1980) The effect of temperature on the formation of sheathed flagella by *Pseudomonas stizolobii. Journal of General Microbiology* **117**, 111–17.

Gardan, L., Morando, B. & Guinebretiere, J.P. (1984) The use of a micromethod of auxanogram and data analysis for identification of Pseudomonads. *Proceedings of the 2nd Working Group on Pseudomonas syringae pathovars. Sounion, Greece, April 1984*, pp. 37–9. (Ed C.G. Panagopoulos, P.G. Psallidas & A.S. Alivizatos). Hellenic Phytopathological Society, Athens.

Garrett, C.M.E. (1987) *Pseudomonas syringae.* In *Handbook of European Plant Diseases,*

(Ed I.M. Smith). Blackwell Scientific Publications, Oxford. (In press).

Garrett, C.M.E., Panagopoulos, C.G. & Crosse, J.E. (1966) Comparison of plant pathogenic pseudomonads from fruit trees. *Journal of Applied Bacteriology* **29**, 342–56.

Gasperini, C., Mazzucchi, U. & Bertolini, M. (1982) Casi di avvizzimento batterico da *Erwinia stewartii* in pianura padana. *L'Informatore Agrario* **48**, 23513–16.

Glynn, L.E. & Steward, M.W. (1977) *Immunochemistry: An Advanced Textbook*. John Wiley & Sons, Chichester.

Goldsworthy, M.C. (1926) Studies on the spot disease of cauliflower; a use of serum diagnosis. *Phytopathology* **16**, 877–84.

Gordon, R.E. & Rynearson, T.K. (1963) Maintenance of strains of Bacillus species. In *Culture Collections: Perspectives and Problems*. (Ed S.M. Martin). University Press, Toronto.

Goto, M. & Starr, M.P. (1971) A comparative study of *Pseudomonas andropogonis, P. stizolobii*, and *P. alboprecipitans*. *Annales of the Phytopathological Society of Japan* **37**, 233–41.

Grieff, D (1969) Safe procedure for opening evacuated glass ampoules containing dried pathogens. *Applied Microbiology* **18**, 130.

Grogan, R.G., Misaghi, I.J., Kimble, K.A., Greathead, A.S., Ririe, D. & Bardin, R. (1977) Varnish spot, a destructive disease of lettuce in California caused by *Pseudomonas cichorii*. *Phytopathology* **67**, 957–60.

Gupta, P. & Swarup, G. (1968) On the ear-cockle and yellow ear rot diseases of wheat. 1. Symptoms and histopathology. *Indian Phytopathology* **22**, 318–22.

Hale, C.N. & Close, R.C. (1974) A survey of the occurrence of bacterial wilt of lucerne in New Zealand. *New Zealand Journal of Experimental Agriculture* **2**, 75–7.

Hattingh, M.J. & Walters, D.F. (1981) Stalk and leaf necrosis of onion caused by *Erwinia herbicola*. *Plant Disease* **65**, 615–18.

Hayward, A.C. (1964) Characteristics of *Pseudomonas solanacearum*. *Journal of Applied Bacteriology* **27**, 265–77.

Hayward, A.C. & Waterston, J.M. (1964a) *Xanthomonas citri*. *C.M.I. Descriptions of Pathogenic Fungi and Bacteria* **11**. Commonwealth Mycological Institute, Kew.

Hayward, A.C. & Waterston, J.M. (1964b) *Xanthomonas malvacearum*. *C.M.I. Descriptions of Pathogenic Fungi and Bacteria* **12**. Commonwealth Mycological Institute, Kew.

Hayward, A.C. & Waterston, J.M. (1964c) *Corynebacterium insidiosum*. *C.M.I. Descriptions of Pathogenic Fungi and Bacteria* **13**. Commonwealth Mycological Institute, Kew.

Hayward, A.C. & Waterston, J.M. (1964d) *Corynebacterium sepedonicum*. *C.M.I. Descriptions of Pathogenic Fungi and Bacteria* **14**. Commonwealth Mycological Institute, Kew.

Hayward, A.C. & Waterston, J.M. (1964e) *Corynebacterium michiganense*. *C.M.I. Descriptions of Pathogenic Fungi and Bacteria* **19**. Commonwealth Mycological Institute, Kew.

Hayward, A.C. & Waterston, J.M. (1964f) *Xanthomonas vesicatoria*. *C.M.I. Descriptions of Pathogenic Fungi and Bacteria* **20**. Commonwealth Mycological Institute, Kew.

Hayward, A.C. & Waterston, J.M. (1965a) *Corynebacterium flaccumfaciens*. *C.M.I. Descriptions of Pathogenic Fungi and Bacteria* **43**. Commonwealth Mycological Institute, Kew.

Hayward, A.C. & Waterston, J.M. (1965b) *Erwinia amylovora*. *C.M.I. Descriptions of Pathogenic Fungi and Bacteria* **44**. Commonwealth Mycological Institute, Kew.

Hayward, A.C. & Waterston, J.M. (1965c) *Pseudomonas phaseolicola*. *C.M.I. Descriptions of Pathogenic Fungi and Bacteria* **45**. Commonwealth Mycological Institute, Kew.

Hayward, A.C. & Waterston, J.M. (1965d) *Xanthomonas campestris. C.M.I. Descriptions of Pathogenic Fungi and Bacteria* **47**. Commonwealth Mycological Institute, Kew.

Hayward, A.C. & Waterston, J.M. (1965e) *Xanthomonas phaseoli. C.M.I. Descriptions of Pathogenic Fungi and Bacteria* **48**. Commonwealth Mycological Institute, Kew.

Hayward, A.C. & Waterston, J.M. (1965f) *Xanthomonas pruni. C.M.I. Descriptions of Pathogenic Fungi and Bacteria* **50**. Commonwealth Mycological Institute, Kew.

Hedges, F. (1924) A study of bacterial pustule of soybean and a comparison of *Bact. phaseoli sojense* Hedges with *Bact. phaseoli* E.F.S. *Journal of Agricultural Research* **29**, 229–51.

Hedges, F. (1926) Bacterial wilt of beans (*Bacterium flaccumfaciens* Hedges), including comparisons with Bacterium phaseoli. *Phytopathology* **16**, 1–22.

Hellmers, G. (1958) Four wilt diseases of perpetual flowering carnation in Denmark. *Dansk Botanisk Arkiv* **18**, 1–200.

Hoppe, P.E. & Kelman, A. (1969) Bacterial top and stalk rot disease of corn in Wisconsin. *Plant Disease Reporter* **53**, 66–70.

Hugh, R. & Leifson, E. (1953) The taxonomic significance of fermentative versus oxidative metabolism of carbohydrates by various Gram-negative bacteria. *Journal of Bacteriology* **66**, 24–6.

Hughes, C.G. (1961) Gumming disease. In *Sugar-Cane Diseases of the World*, Vol. I, pp. 55–71. (Ed J.P. Martin, E.V. Abbott & C.G. Hughes). Elsevier, Amsterdam.

Janse, J.D. (1981a) The bacterial disease of ash (*Fraxinus excelsior*), caused by *Pseudomonas syringae* subsp. *savastanoi* pv. *fraxini* I. History, occurrence and symptoms. *European Journal of Forest Pathology* **11**, 306–15.

Janse, J.D. (1981b) The bacterial disease of ash (*Fraxinus excelsior*), caused by *Pseudomonas syringae* subsp. *savastanoi* pv. *fraxini* II. Etiology and taxonomic considerations. *European Journal of Forest Pathology* **11**, 425–32.

Janse, J.D. (1982) *Pseudomonas syringae* subsp. *savastanoi* (ex Smith) subsp. nov., nom. rev., the bacterium causing excrescences on *Oleaceae* and *Nerium oleander* L. *International Journal of Systematic Bacteriology* **32**, 166–9.

Janse, J.D. & Miller, H.J. (1983) Yellow disease in *Scilla tubergeniana* and related bulbs caused by *Xanthomonas campestris* pv. *hyacinthi*. *Netherlands Journal of Plant Pathology* **89**, 203–6.

Jones, J.B., Jones, J.P., McCarter, S.M. & Stall, R.E. (1984) *Pseudomonas viridiflava*: Causal agent of bacterial leaf blight of tomato. *Plant Disease* **68**, 341–2.

Kelman, A. (1953) The bacterial wilt caused by *Pseudomonas solanacearum*. *North Carolina Agricultural Experiment Station Technical Bulletin* **99**.

Kelman, A. (1954) The relationship of pathogenicity in *Pseudomonas solanacearum* to colony appearance on a tetrazolium medium. *Phytopathology* **44**, 693–5.

Kennedy, B.W. & King, T.H. (1962) Angular leaf spot of strawberry caused by *Xanthomonas fragariae* sp. nov. *Phytopathology* **52**, 873–5.

Kersters, K. & De Ley, J. (1984) *Agrobacterium* Conn 1942. In *Bergey's Manual of Systematic Bacteriology* Vol. I, pp. 244–54, (Ed N.R. Krieg & R.M. Holt). Williams & Wilkins, Baltimore.

Keyworth, W.G. & Howell, S.J. (1961) Studies on the silvering disease of red beet. *Annals of Applied Biology* **49**, 173–95.

Keyworth, W.G., Howell, S.J. & Dowson, W.J. (1956) *Corynebacterium betae* (sp. nov.) the causal organism of silvering disease of red beet. *Plant Pathology* **5**, 88.

King, E.O., Ward, M.K. & Raney, D.E. (1954) Two simple media for the demonstration of pyocyanin and fluorescein. *Journal of Laboratory and Clinical Medicine* **44**, 301–7.

Kivilaan, A. & Scheffer, R.P. (1958) Factors affecting development of bacterial stem rot of pelargonium. *Phytopathology* **48**, 185–91.

Kovacs, N. (1956) Identification of *Pseudomonas pyocyanea* by the oxidase reaction. *Nature, London* **178**, 703.

Krieg, N.R. & Holt, J.G. (Eds.) (1984) *Bergey's Manual of Systematic Bacteriology* Vol. I, Williams & Wilkins, Baltimore.

Lelliott, R.A. (1956) Slow wilt of carnation caused by a species of *Erwinia*. *Plant Pathology* **5**, 9–23.

Lelliott, R.A. (1965) The preservation of plant pathogenic bacteria. *Journal of Applied Bacteriology* **28**, 181–93.

Lelliott, R.A. (1968) The diagnosis of fireblight (*Erwinia amylovora*) and some diseases caused by *Pseudomonas syringae*. *EPPO Publications, Series A* **45**, 27–34.

Lelliott, R.A. (1979) Lucerne Bacterial Wilt. *ADAS Leaflet* **651**. Ministry of Agriculture, Fisheries and Food (Publications) UK.

Lelliott, R.A. (1987) *Corynebacterium insidiosum, Corynebacterium sepedonicum, Erwinia amylovora, Erwinia caratovora, Pseudomonas viridiflava*. In *Handbook of European Plant Diseases*. (Ed I.M. Smith). Blackwell Scientific Publications, Oxford. (In press).

Lelliott, R.A. & Billing, E. (1984) Fireblight of Apple and Pear. *ADAS Leaflet* **571**. Ministry of Agriculture, Fisheries and Food (Publications) UK.

Lelliott, R.A., Billing, E. & Hayward, A.C. (1966) A determinative scheme for the fluorescent plant pathogenic pseudomonads. *Journal of Applied Bacteriology* **29**, 470–89.

Lelliott, R.A. & Sellar, P.W. (1976) The detection of latent ring rot (*Corynebacterium sepedonicum* (Spieck. *et* Koth.) Skapt. *et* Burkh.) in potato stocks. *EPPO Bulletin* **6**, 101–6.

Locke, T. & Barnes, D. (1979) *Xanthomonas corylina* on cobnuts and filberts. *Plant Pathology* **28**, 53.

Lozano, J.C. (1975) Bacterial blight of cassava. *PANS (Pest Articles and News Summaries)* **21**, 38–43.

Lozano, J.C. (1977) Identification of bacterial leaf blight in rice, caused by *Xanthomonas oryzae*, in America. *Plant Disease Reporter* **61**, 644–8.

Lozano, J.C. & Sequira, L. (1970) Differentiation of races of *Pseudomonas solanacearum* by a leaf infiltration technique. *Phytopathology* **60**, 838–8.

Lund, B.M. (1969) Properties of some pectolytic, yellow-pigmented, Gram negative bacteria isolated from fresh cauliflowers. *Journal of Applied Bacteriology* **32**, 60–7.

Lund, B.M. (1972) Isolation of pectolytic clostridia from potatoes. *Journal of Applied Bacteriology* **35**, 609–14.

Lund, B.M. (1979) Bacterial soft-rot of potatoes. In *Plant Pathogens*, pp. 19–49. (Ed D.W. Lovelock). Academic Press, London.

McCulloch, L. (1925) *Aplanobacter insidiosum* n.sp., the cause of an alfalfa disease. *Phytopathology* **15**, 496–7.

McLean, D.M. (1958) A seed-borne bacterial cotyledon spot of squash. *Plant Disease Reporter* **42**, 425–6.

McPherson, G.M., Bradbury, J.F. & Preece, T.F. (1977) *Xanthomonas pelargonii. C.M.I. Descriptions of Pathogenic Fungi and Bacteria* **560**. Commonwealth Mycological Institute, Kew.

Mackie, T.J. & McCartney, J.E. (1978) *Medical Microbiology*, Vol. I, 13th Edn. Churchill Livingstone, Edinburgh.

Maraite, H. & Weyns, J. (1979) Distinctive physiological, biochemical and pathogenic characteristics of *Xanthomonas manihotis* and *X. cassavae*. In *Diseases of Tropical Food Crops*, pp. 103–17. (Ed H. Maraite & J.A. Meyer). Université Catholique de Louvain-la-Neuve, Belgium.

Martin, J.P. & Robinson, P.E. (1961) Leaf scald. In *Sugar-Cane Diseases of the World*, Vol. I, pp. 78–107. (Ed J.P. Martin, E.V. Abbott & C.G. Hughes.) Elsevier, Amsterdam.

Martin, J.P. & Wismer, C.A. (1961) Red stripe. In *Sugar-Cane Diseases of the World*, Vol. I, pp. 109–26. (Ed J.P. Martin, E.V. Abbott & C.G. Hughes). Elsevier, Amsterdam.

Mazzucchi, U., Bazzi, C. & Calzolari, A. (1983) Diagnosis del marciume anulare della patata. *Informatore Fitopatologico* **5**, 26–36.

Mazzucchi, U. & Dalli, A. (1975) *Xanthomonas phaseoli* var. *fuscans* agente di una nuova batteriosi del fagioli in Emilia: il cancro batterico. *Informatore Fitopatologico* **25**, 28–34.

Melling, J. & Allner, K. (1981) The containment of micro-organisms. In *Essays in Applied Microbiology*. (Ed J.R. Norris & M.H. Richmond). John Wiley & Sons, Chichester.

Mergaert, J., Verdonck, L., Kersters, K., Swings, J., Boeufgras, J.M. & De Ley, J. (1984) Numerical taxonomy of *Erwinia* species using API systems. *Journal of General Microbiology* **130**, 1893–910.

Miller, P.R. & Pollard, H.L. (1976) *Multilingual Compendium of Plant Diseases*. American Phytopathological Society, Minnesota.

Miller, P.W., Bollen, W.B., Simmons, J.E., Gross, H.N. & Barss, H.P. (1940) The pathogen of filbert bacteriosis compared with *Phytomonas juglandis*, the cause of walnut blight. *Phytopathology* **30**, 713–33.

Mingtan, L. & Watson, T. (1973) Bacterial spot of tomato and pepper in California. *Plant Disease Reporter* **57**, 258–9.

Moffett, M.L. (1973) Bacterial spot of stone fruit in Queensland. *Australian Journal of Biological Science* **26**, 171–9.

Moffett, M.L., Peterson, R.A. & Wood, B.A. (1979) Bacterial black spot of mango. *Australian Plant Pathology* **8**, 54–6.

Mulrean, E.N. & Schroth, M.N. (1982) Ecology of *Xanthomonas campestris* pv. *juglandis* on Persian (English) walnuts. *Phytopathology* **72**, 434–8.

Munnecke, D.E. Chandler, P.A. & Starr, M.P. (1963) Hairy root (*Agrobacterium rhizogenes*) of field roses. *Phytopathology* **53**, 788–99.

New, P.B. & Kerr, A. (1971) A selective medium for *Agrobacterium radiobacter* biotype 2. *Journal of Applied Bacteriology* **34**, 233–6.

Nishiyama, K., Kusaba, T., Ohta, K., Nahata, K. & Ezuka, A. (1979) Bacterial black rot of tulip caused by *Pseudomonas andropogonis*. *Annales of the Phytopathological Society of Japan* **45**, 668–74.

Ostle, A.G. & Holt, J.G. (1982) Nile Blue A as fluorescent stain for poly-β-hydroxybutyrate. *Applied and Environmental Microbiology* **44**, 238–41.

Ou, S. (1972) *Rice Diseases*. Commonwealth Mycological Institute, Kew.

Palleroni, N.J. (1984) *Pseudomonas* Migula 1894. In *Bergey's Manual of Systematic Bacteriology*, Vol. I, pp. 141–99. (Ed N.R. Krieg & J.G. Holt). Williams & Wilkins, Baltimore.

Panagopoulos, C.G. (1987) *Agrobacterium tumefaciens*, *Xanthomonas ampelina*. In *Handbook of European Plant Diseases*. (Ed I.M. Smith). Blackwell Scientific Publications, Oxford. (In press).

Pathak, K.N. & Swarup, G. (1984) Incidence of *Corynebacterium michiganense* pv. *tritici* in the ear-cockle nematode (*Anguina tritici*), galls and pathogenicity. *Indian Phytopathology* **37**, 267–70.

Paton, A.M. (1958) Pectin-decomposing strains of *Pseudomonas*. *Nature, London* **181**, 61–2.

Pepper, E.H. (1967) Stewart's wilt of corn. *American Phytopathological Society Monograph No. 4*. St. Paul, Minnesota, USA.

Perombelon, M.C.M. & Kelman, A. (1980) Ecology of the soft-rot erwinias. *Annual Review of Phytopathology* **18**, 361–87.

Persley, G.J. (1977) Distribution and importance of cassava bacterial blight in Africa. In *Cassava Bacterial Blight, Report of an interdisciplinary workshop held at IITA, Ibadan, Nigeria, 1–4 November, 1976*, pp. 9–14. (Ed G.J. Persley, E.R. Terry & R. MacIntyre) International Development Research Centre, Ottowa.

Pohronezney, K., Larsen, P. & Leben, C. (1978) Observations on cucumber fruit invasion by *Pseudomonas lachrymans*. *Plant Disease Reporter* **62**, 306–9.

Preece, T.F., Wong, W.C. & Adegaye, A.O. (1979) Diagnosis of watermark in willows and some characteristics of *Erwinia salicis* (Day) Chester. In *Plant Pathogens* pp. 1–17. (Ed D.W. Lovelock). Academic Press, London.

Renfro, B.L. & Sprague, E.W. (1959) Reaction of *Medicago* species to eight Alfalfa pathogens. *Agronomy Journal* **51**, 481–3.

Ridé, M. (1987) *Xanthomonas populi*. In *Handbook of European Plant Diseases*. (Ed I.M. Smith). Blackwell Scientific Publications, Oxford. (In press).

Ridé, M. & Marcelin, H. (1983) La nécrose bactérienne de la Vigne: *Xanthomonas ampelina*. *Bulletin Technique des Pyrenées-Orientales* **106**.

Ridé, M. & Ridé, S. (1978) *Xanthomonas populi*, (Ridé) Comb. nov. (syn. *Aplanobacter populi* Ridé), specificité, variabilité et absence de relations avec *Erwinia cancerogena European Journal of Forest Pathology* **8**, 310–33.

Ridé, M., Ridé, S. & Poutier, J.C. (1978) Factors affecting the survival of *Xanthomonas populi* on aerial structures of poplar. *Proceedings of the 4th International Conference on Plant Pathogenic Bacteria, Angers 1978*, pp. 803–14. Station de Pathologic Végétale et Phytobacteriologic INRA, Angers.

Robbs, C.F., Neto, J.R., Ribeiro, R. de L.D. & Kimura, O. (1981) Bacterial leaf spot of ornamental *Triplaris* caused by *Pseudomonas andropogonis*. *Proceedings of the 5th International Conference on Plant Pathogenic Bacteria, Cali, Colombia, 1981*, pp. 54–9. (Ed J.C. Lozano). Centro International de Agricultura Tropical, Cali, Colombia.

Robert, A.L. (1967) Bacterial wilt and Stewart's leaf blight of corn. *Farmers' Bulletin US Department of Agriculture* **2092**.

Roberts, P. (1974) *Erwinia rhapontici* (Millard) Burkholder associated with pink grain of wheat. *Journal of Applied Bacteriology* **37**, 353–8.

Roitt, I.M. (1984) *Essential Immunology*, 5th Edn. Blackwell Scientific Publications, Oxford.

Saaltink, G.J. & Kamerman, W. (1971) De geelziekte van hyacint., **34**. Bulb Research Centre Praktijkmeded-elhigen, Lisse.

Saaltink, G.J. & Maas Geesteranus, H.P. (1969) A new disease in Tulip caused by *Corynebacterium oortii* nov. spec. *Netherlands Journal of Plant Pathology* **75**, 123–8.

Samson, R. & Nassan-Agha, N. (1978) Biovars and serovars among 129 strains of *Erwinia chrysanthemi*. *Proceedings of the 4th International Conference on Plant Pathogenic Bacteria, Angers, 1978*, pp. 547–53. Station de Pathologic Végétale et Phytobacteriologic INRA, Angers.

Sasser, M. & Miller, L.T. (1984) Identification of Pseudomonads by fatty acid profiling. *Proceedings of the 2nd Working Group on Pseudomonas syringae pathovars*. Sounion, Greece, April 1984, pp. 45–6.

Scarlett, C.M. Fletcher, J.T., Roberts, P. & Lelliott, R.A. (1978) Tomato pith necrosis caused by *Pseudomonas corrugata* n. sp. *Annals of Applied Biology* **88**, 105–14.

Schaad, N.W. (1979) Serological identification of plant pathogenic bacteria. *Annual Review of Phytopathology* **17**, 123–47.
Schaad, N.W., Kado, C.I. & Sumner, D.R. (1975) Synonymy of *Pseudomonas avenae* Manns 1909 and *Pseudomonas alboprecipitans* Rosen 1922. *International Journal of Systematic Bacteriology* **25**, 133–7.
Schroth, M.N., Thompson, J.P. & Hildebrand, D.C. (1965) Isolation of *Agrobacterium tumefaciens — A. radiobacter* group from the soil. *Phytopathology* **55**, 645–7.
Sellwood, J.E. & Lelliott, R.A. (1978) Internal browning of hyacinth caused by *Erwinia rhapontici*. *Plant Pathology* **27**, 120–4.
Sinclair, J.B. & Shurtleff, M.C. (1975) *A Compendium of Soybean Diseases*. American Phytopathological Society, St Paul, Minnesota.
Skerman, V.B.D., McGowan, V. & Sneath, P.H.A. (1980) Approved lists of bacterial names. *International Journal of Systematic Bacteriology* **30**, 225–420.
Smith, E.F. (1905) *Bacteria in Relation to Plant Diseases*, Vol. 1. Carnegie Institute of Washington, Washington DC.
Smith, E.F. (1911) *Bacteria in Relation to Plant Diseases*, Vol. 2. Carnegie Institute of Washington, Washington DC.
Smith, E.F. (1914) *Bacteria in Relation to Plant Diseases*, Vol. 3. Carnegie Institute of Washington, Washington DC.
Smith, E.F. (1920) *An Introduction to Bacterial Diseases of Plants*. W.B. Saunders, Philadelphia.
Smith, E.F., Jones, L.R. & Reddy, C.S. (1919) The black chaff of wheat. *Science (Washington)* **50**, 48.
Smith, M.A. & Ramsey, G.B. (1953) Bacterial spot of broccoli. *Phytopathology* **43**, 583–4.
Snieszko, S.F. & Bonde, R. (1943) Studies on the morphology, physiology, serology, longevity, and pathogenicity of *Corynebacterium sepedonicum*. *Phytopathology* **33**, 1032–44.
Stall, R.E. & Seymour, C.P. (1983) Canker, a threat to Citrus in the Gulf-Coast States. *Plant Disease* **67**, 581–5.
Stanier, R.Y., Palleroni, N.J. & Doudoroff, M. (1966) The aerobic pseudomonads: a taxonomic study. *Journal of General Microbiology* **43**, 159–271.
Stapp, C. (1961) *Bacterial Plant Pathogens*. Oxford University Press, Oxford.
Starr, M.P. & Pirone, P.P. (1942) *Phytomonas poinsettiae* n. sp., the cause of a bacterial disease of poinsettia. *Phytopathology* **32**, 1076–81.
Stead, D.E. (1974) *Soft rotting of potatoes by the blackleg bacterium, Erwinia atroseptica* (PhD thesis, London University).
Steyn, P.L., Viljoen, N.M. & Kotz, J.M. (1974) The casual organism of bacterial black spot of mangoes. *Phytopathology* **64**, 1400–4.
Strider, D.L. (1969) Bacterial canker of tomato caused by *Corynebacterium michiganense*. *North Carolina Agricultural Experiment Station Technical Bulletin* **193**.
Suslow, T.V. & Schroth, M.N. (1982) Role of deleterious rhizobacteria as minor pathogens in reducing crop growth. *Phtyopathology* **72**, 111–15.
Sykes, G. (1967) *Disinfection and Sterilisation*. J.B. Lippincott Co, Philadelphia.
Taylor, E.H., Bradbury, J.F. & Preece, T.F. (1981) *Xanthomonas campestris* pv. *begoniae*. *C.M.I. Descriptions of Pathogenic Fungi and Bacteria* **699**. Commonwealth Mycological Institute, Kew.
Taylor, J.D. (1972) Races of *Pseudomonas pisi* and sources of resistance in field and garden peas. *New Zealand Journal of Agricultural Research* **15**, 441–7.
Taylor, J.D., Dudley, C.L. & Presly, L. (1979) Studies of halo blight seed infection and disease transmission in dwarf beans. *Annals of Applied Biology* **93**, 267–77.

Taylor, J.D. & Dye, D.W. (1972) A survey of the organisms associated with bacterial blight of peas. *New Zealand Journal of Agricultural Research* **15**, 432–40.

Tesoriero, L.A., Fahy, P.C. & Gunn, L.V. (1982) First record of a bacterial rot of onion in Australia caused by *Pseudomonas gladioli* pv. *allicola* and association with internal browning caused by *Pseudomonas aeruginosa*. *Australian Plant Pathology* **11**, 56–7.

Thomson, S.V., Hildebrand, D.C. & Schroth, M.N. (1981) Identification and nutritional differentiation of the *Erwinia* sugar beet pathogen from members of *Erwinia carotovora* and *Erwinia chrysanthemi*. *Phytopathology* **71**, 1037–42.

Thornley, M.J. (1960) The differentiation of *Pseudomonas* from other Gram-negative bacteria on the basis of arginine metabolism. *Journal of Applied Bacteriology*, **23**, 37–52.

Torrance, L. & Jones, R.A.C. (1981) Recent developments in serological methods suited for use in routine testing for plant viruses. *Plant Pathology* **30**, 1–24.

US Public Health Service (1977) *Laboratory Safety at the Center for Disease Control*, Department of Health, Education and Welfare, Publication No. CDC 77–8118. Center for Disease Control, Atlanta.

Van den Mooter, M., Swings, J., De Cleene, M., Leyns, F. & De Ley, J. (1981) Isolation and identification of *Xanthomonas campestris* pv. *graminis* from forage grasses in Belgium. *Parasitica* **37**, 23–8.

Van der Zwet, T. & Keil, H.L. (1979) *Fireblight. A Bacterial Disease of Rosaceous Plants*. USDA Agriculture Handbook, No. 510.

Van Regenmortel, M.H.V. (1982) *Serology and Immunochemistry of Plant Viruses*. Academic Press, London.

Volcani, Z., Zutra, D. & Cohn, R. (1970) A new leaf and fruit spot disease of pepper caused by *Corynebacterium michiganense*. *Plant Disease Reporter* **54**, 804–6.

Weir, D.M. (1978) *Handbook of Experimental Immunology*. Blackwell Scientific Publications, Oxford.

Wilkie, J.P. (1972) Halo blight of oats in New Zealand. *New Zealand Journal of Agricultural Research* **15**, 461–81.

Wilkie, J.P. (1973). Basal glume rot of wheat in New Zealand. *New Zealand Journal of Agricultural Research* **16**, 155–60.

Wilkie, J.P. & Dye, D.W. (1974a) *Pseudomonas cichorii* causing tomato and celery diseases in New Zealand. *New Zealand Journal of Agricultural Research* **17**, 123–30.

Wilkie, J.P., & Dye, D.W. (1974b) *Pseudomonas tomato* in New Zealand. *New Zealand Journal of Agricultural Research* **17**, 131–5.

Wilkie, J.P., Dye, D.W. & Watson, D.R.W. (1973) Further hosts of *Pseudomonas viridiflava*. *New Zealand Journal of Agricultural Research* **16**, 315–23.

Williams, P.H. & Keen, N.T. (1966) Bacterial blight of radish. *Plant Disease Reporter* **50**, 192–5.

Wimalajeewa, D.L.S. & Washington, W.S. (1980) Bacterial blight of hazelnut. *Australian Plant Pathology* **9**, 113–14.

Wong, W.C. & Preece, T.F. (1979) Identification of *Pseudomonas tolaasii*: the white line in agar and mushroom block rapid pitting tests. *Journal of Applied Bacteriology* **47**, 401–7.

Wong, W.C., Fletcher, J.T., Unsworth, B.A. & Preece, T.F. (1982) A note on ginger blotch, a new bacterial disease of the cultivated mushroom, *Agaricus bisporus*. *Journal of Applied Bacteriology* **52**, 43–8.

Wormald, H. (1924) The Mulberry 'Blight' in Britain. *Annals of Applied Biology* **11**, 169–74.

Wormald, H. (1939) *Diseases of Fruit and Hops*, p. 245. Crosby Lockwood & Son Ltd, London.

Wormald, H. (1945) Physiologic races of the crown gall organism in Britain. *Transactions of the British Mycological Society* **28**, 134.

Young, J.M. (1970) Drippy gill: a bacterial disease of cultivated mushrooms caused by *Pseudomonas agarici* n.sp. *New Zealand Journal of Agricultural Research* **13**, 977–90.

Young, J.M., Dye, D.W., Bradbury, J.F., Panagopoulos, C.G. & Robbs, C.F. (1978) A proposed nomenclature and classification for plant pathogenic bacteria. *New Zealand Journal of Agricultural Research* **21**, 153–77.

Zaumeyer, W.J. (1932) Comparative pathological histology of three bacterial diseases of bean. *Journal of Agricultural Research* **44**, 605–32.

Zaumeyer, W.J. & Thomas, H.R. (1957) A monographic study of bean diseases and methods for their control. *Technical Bulletin, USDA* **868**, 733–74.

Index

Diseases are not indexed by common name but references to them in the text can be found via the Latin binomials of their causal pathogens. A host–pathogen index can be found on pp. 22–36.

Methods in Plant Pathology

SERIES EDITOR

T.F. PREECE

BSc MSc DSc FIBiol AIMLs

University of Leeds

1 Methods in Plant Virology

S.A. HILL

14-50

METHODS FOR THE DIAGNOSIS
OF BACTERIAL DISEASES OF PLANTS